T0238129

Lecture Notes in Artificial Intelligence   6036

Edited by R. Goebel, J. Siekmann, and W. Wahlster

Subseries of Lecture Notes in Computer Science

Enrico Francesconi   Simonetta Montemagni
Wim Peters   Daniela Tiscornia (Eds.)

Semantic Processing of Legal Texts

Where the Language of Law
Meets the Law of Language

 Springer

Series Editors

Randy Goebel, University of Alberta, Edmonton, Canada
Jörg Siekmann, University of Saarland, Saarbrücken, Germany
Wolfgang Wahlster, DFKI and University of Saarland, Saarbrücken, Germany

Volume Editors

Enrico Francesconi
Daniela Tiscornia
Institute of Legal Information, Theory and Techniques, ITTIG-CNR
Via dei Barucci 20, 50127 Florence, Italy
E-mail: {francesconi, tiscornia}@ittig.cnr.it

Simonetta Montemagni
Istituto di Linguistica Computazionale "Antonio Zampolli" (ILC) - CNR
Area della Ricerca di Pisa, Via Moruzzi 1, 56124 Pisa, Italy
E-mail: simonetta.montemagni@ilc.cnr.it

Wim Peters
University of Sheffield, Department of Computer Science
Regent Court, 211 Portobello Street, Sheffield S1 4DP, UK
E-mail: w.peters@dcs.shef.ac.uk

Library of Congress Control Number: 2010295554

CR Subject Classification (1998): I.2, H.3, H.4, H.2.8, H.2, J.1

LNCS Sublibrary: SL 7 – Artificial Intelligence

ISSN 0302-9743

ISBN 978-3-642-12836-3 Springer Berlin Heidelberg New York

This work is subject to copyright. All rights are reserved, whether the whole or part of the material is
concerned, specifically the rights of translation, reprinting, re-use of illustrations, recitation, broadcasting,
reproduction on microfilms or in any other way, and storage in data banks. Duplication of this publication
or parts thereof is permitted only under the provisions of the German Copyright Law of September 9, 1965,
in its current version, and permission for use must always be obtained from Springer. Violations are liable
to prosecution under the German Copyright Law.

springer.com

© Springer-Verlag Berlin Heidelberg 2010

Typesetting: Camera-ready by author, data conversion by Scientific Publishing Services, Chennai, India
Printed on acid-free paper   06/3180

Preface

The legal domain represents a primary candidate for Web-based information distribution, exchange and management, as testified by the numerous e-government, e-justice and e-democracy initiatives worldwide. The last few years have seen a growing body of research and practice in the field of artificial intelligence and law addressing aspects such as automated legal reasoning and argumentation, semantic and cross-language legal information retrieval, document classification, legal drafting, legal knowledge discovery and extraction. Many efforts have also been devoted to the construction of legal ontologies and their application to the law domain.

A number of different workshops and conferences have been organized on these topics in the framework of the artificial intelligence and law community: among them, the ICAIL (International Conference on Artificial Intelligence and Law) and the Jurix (International Conference on Legal Knowledge and Information Systems) conferences; several workshops on legal ontologies have been held by the AI&Law Association (LOAIT) and by the Legal XML Community (LegalXML Workshops and Legal XML Summer School). In all these events, the topics of language resources and human language technologies receive increasing attention.

The situation is quite different within the computational linguistics community, where little attention has been paid to the legal domain besides a few isolated contributions and/or projects focussing on the processing of legal texts. In this context, the editors of this book organized a Workshop on "Semantic Processing of Legal Texts," which was held in Marrakech (Morocco) in 2008, in the framework of the 6th International Conference on Language Resources and Evaluation (LREC–2008). The workshop offered the possibility for the two communities to meet, exchange information, compare perspectives and share experiences and concerns on the topic of legal knowledge extraction and management. Both research communities can benefit from this interaction: the legal artificial intelligence community can gain insight into state–of–the–art linguistic technologies, tools and resources, and the computational linguists can take advantage of the large and often multilingual legal resources – corpora as well as lexicons and ontologies – for training and evaluation of current Natural Language Processing (NLP) technologies and tools. The main focus of the workshop was the automatic extraction of relevant information from legal texts and the structured organization of this extracted knowledge for legal knowledge representation and scholarly activity, with particular emphasis on the crucial role played by language resources and human language technologies.

The number of received submissions, the variety of perspectives and approaches witnessed by the accepted papers, the number and variety of participants from 17 different countries all over the world, both from the academic and industrial communities and with different (i.e., legal, linguistic as well as computational) backgrounds, as well as the stimulating discussion both speakers and attendees engaged in during the workshop indicate this as a promising field, which combines legal informatics and natural language processing in innovative and productive ways. This fact persuaded us to invest some time to wrap up the current debate in a book including the revised and expanded versions of selected papers presented at the workshop which were complemented with invited contributions of leading researchers and groups eminently active in the field.

The present volume is thus the outcome of this joint effort, covering some of the main issues in current research on semantic processing of legal texts, as well as providing new exciting avenues for the years to come. The papers report research from the academic and industrial communities, from different countries dealing with different languages, including less–resourced ones, and all together provide an articulated picture of the current achievements and challenges in the field. We are aware that several books have been published in the legal artificial intelligence community on topics such as legal standards and legal ontologies[1]. Up to now, however, no comprehensive overview of the field of semantic processing of legal texts exists, combining views and perspectives from the computational linguistic and legal communities. The volume intends to fill this gap, by enabling readers to gain insight into state–of–the–art NLP technologies, tools and the availability of legal resources, in the form of corpora, lexicons and ontologies. In particular, we hope that it will be seminal for the future integration of NLP techniques into recently established approaches to legal analysis such as formal ontology design and acquisition, and contrastive analysis of legal systems. In the book the challenges of the field are tackled from different perspectives, thus providing the reader with an overview of the current debate.

The book is organized into thematic sections, each covering core topics addressing, from different perspectives, the complex relation between legal text and legal knowledge. The boundaries of these sections are of course not strict, but delineate general aspects of thematic segmentation.

Part 1, "Legal Text Processing and Information Extraction," focuses on the analysis of legal texts through a variety of NLP techniques, with a specific view to information extraction tasks. The first paper of this section, by Venturi, investigates the peculiarities of legal language with respect to ordinary language for

---

[1] Benjamins R., Casanovas P., Gangemi A. (eds.), Law and the Semantic Web, Springer, 2005; Ajani G.M., Peruginelli G., Sartor G. and Tiscornia D. (eds.), The multilingual Complexity of European Law, European Press Academic Publishing, 2006; Biagioli C., Francesconi E., Sartor G. (eds.), Proceedings of the V Legislative XML Workshop, European Press Academic Publishing, 2007; Francesconi E.,Technologies for European Integration, European Press Academic Publishing, 2007; Breuker J., Casanovas P., Klein M., Francesconi E. (eds.), Law, Ontologies and the Semantic Web, IOS Press 2008; Casanovas P., Sartor G., Casellas N., Rubino R. (eds.), Computable Models of the Law, LNAI 4884, Spinger 2008.

both Italian and English as a basic prerequisite for the development of domain–specific knowledge management applications. The second and third papers, respectively, by Dozier et al. and by Quaresma and Gonçalves, deal with a basic but still challenging annotation task, i.e., the recognition and classification of legal entities (namely, named entities) mentioned in legal texts. The paper by Wyner et al. closes the section with a survey of recent text–mining approaches used to automatically profile and extract arguments from legal cases.

Part 2, "Legal Text Processing and Construction of Knowledge Resources," illustrates the challenges NLP faces and its achievements in a bootstrapping approach to formal ontology design and construction, based on the integration of ontological and textual resources. In particular, the first paper by Pala et al. illustrates the results of experiments aimed at bootstrapping an electronic dictionary of Czech law terms from texts. The second contribution of this section, by Francesconi et al., presents a methodology for multilingual legal knowledge acquisition and modelling encompassing two complementary strategies; two case-studies combining bottom–up and top-down methodologies for knowledge modelling and learning are presented. The third paper by Bosca and Dini illustrates an ontology induction experiment for individual laws (the Italian "Legge Bassanini" in the case at hand) based on corpora comparison that exploits a domain corpus automatically generated from the Web. In a different vein, the contribution by Ajani et al. illustrates new features recently introduced in the Legal Taxonomy Syllabus (LTS), a tool for building multilingual conceptual dictionaries for the EU law, to tackle the problem of representing the interpretation of terms besides the definitions occurring in the directives, the problem of normative change, and the process of planning legal reforms of European law.

Part 3, "Legal Text Processing and Semantic Indexing, Summarization and Translation," covers a selection of tasks and applications which can benefit from NLP–based functionalities, including legal document indexing (Schweighofer), classification (cfr. de Maat and Winkels; Loza Mencía and Fürnkranz), summarization (Chieze et al.) and translation (Ogawa et al.).

We hope that this example–based overview of semantic processing of legal texts will provide a unique opportunity for researchers operating in the field to gain insight into this active and challenging research area, where the language of law meets the law of language.

We would like to thank all contributing authors, for their enthusiasm, vision and stamina. Last but not least, we intend to express our sincere gratitude to all reviewers for their insightful contributions.

March 2010

Enrico Francesconi
Simonetta Montemagni
Wim Peters
Daniela Tiscornia

Organization

Referees

Danièle Bourcier	Humboldt Universität, Berlin, Germany
Paul Bourgine	CREA, Ecole Polytechnique, Paris, France
Joost Breuker	Leibniz Center for Law, University of Amsterdam, The Netherlands
Pompeu Casanovas	Institut de Dret i Tecnologia, UAB, Barcelona, Spain
Alessandro Lenci	Dipartimento di Linguistica, Università di Pisa, Italy
Leonardo Lesmo	Dipartimento di Informatica, Università di Torino, Turin, Italy
Manfred Pinkal	Department of Computational Linguistics, Saarland University, Germany
Vito Pirrelli	Istituto di Linguistica Computazionale of CNR, Pisa, Italy
Paulo Quaresma	Universidade de Évora, Portugal
Erich Schweighofer	Universität Wien, Rechtswissenschaftliche Fakultät, Vienna, Austria
Tom van Engers	Leibniz Center for Law, University of Amsterdam, The Netherlands
Maria A. Wimmer	Institute for Information Systems, Koblenz University, Germany
Radboud Winkels	Leibniz Center for Law, University of Amsterdam, The Netherlands

Table of Contents

PART I

Legal Text Processing and Information Extraction

Legal Language and Legal Knowledge Management Applications

Giulia Venturi

Institute of Computational Linguistics, CNR
via G. Moruzzi 1, 56124, Pisa, Italy
giulia.venturi@ilc.cnr.it
http://www.ilc.cnr.it

Abstract. This work is an investigation into the peculiarities of legal language with respect to ordinary language. Based on the idea that a shallow parsing approach can help to provide enough detailed linguistic information, this work presents the results obtained by shallow parsing (i.e. chunking) corpora of Italian and English legal texts and comparing them with corpora of ordinary language. In particular, this paper puts the emphasis of how understanding the syntactic and lexical characteristics of this specialised language has practical importance in the development of domain–specific Knowledge Management applications.

Keywords: Parsing Legal Texts, Natural Language Processing, Legal Language, Knowledge Management Applications.

1 Introduction

This work is an investigation into the peculiarities of legal language with respect to ordinary language. Within the specialised–domain field, the increasing need for text processing of large document collections has prompted research efforts devoted to automatically processing the sublanguage used in domain-specific corpora. Interestingly, sublanguage processing encompasses not only the numerous inherent complexities of ordinary Natural Language Processing (NLP), but also the treatment of domain–specific peculiarities. Accordingly, beyond the general NLP difficulties, the specificities of domain–specific features make the automatic processing of these kind of corpora a challenging task and demand specific solutions. This is the reason why this article puts the emphasis of how understanding the characteristics of a specialised language has practical importance in the development of domain–specific Knowledge Management applications.

More than within other research communities, the reciprocal exchange between the bio–medical community and the Natural Language Processing one is a promising example in this direction. The very active research field (see [3] for an updated overview) witnesses that this joint effort between the two is fruitful for the purposes of both communities. It follows that on the one hand a variety of NLP techniques are exploited for a number of domain–specific applications such as Biological Text Mining, the construction of ontological resources, Information Retrieval, and Event Extraction. On the other hand, various efforts are

E. Francesconi et al. (Eds.): Semantic Processing of Legal Texts, LNAI 6036, pp. 3–26, 2010.
© Springer-Verlag Berlin Heidelberg 2010

also devoted to investigating biomedical language characteristics with a view to customise NLP tools, in order to adapt them to the processing of biomedical sub–language.

On the contrary, within the Artificial Intelligence and Law community (AI& Law), to the author's knowledge, little attention has been devoted both to techniques coming from the NLP community, and to research efforts concerned with the analysis of legal language peculiarities. In particular, it has been overlooked how understanding the characteristics of this specialised language can help to shed light on the main difficulties of extracting *semantic information* out of legal documents.

Within the legal domain the situation is made more challenging, with respect to other specialised domains, by the fact that laws are invariably conveyed through natural language. According to linguistic studies, such as Garavelli [20], legal language, although still different from ordinary language, is in fact not dramatically independent from every day speech. It rather makes a *specific use* of lexical and syntactic peculiarities typical of ordinary language. Consequently, it can be seen both as an *extension* and as a *reduction* of the possibilities offered by ordinary language. Moreover, since the 80s the close intertwining between legal knowledge and legal language has been acknowledged within the Artificial Intelligence and Legal Reasoning community. Pointing out some special characteristics of the legal domain, Dyer in Rissland [26] claims that "[m]odeling what a lawyer does is more complex than modeling experts in technical/scientific domains. First, all of these complex conceptualizations are expressed in natural language, so modeling the comprehension ability of the lawyer requires solving the natural language problem". However, unfortunately, as it will be discussed in Section 2, few research activities have focussed on this topic.

According to these premises, this article aims at suggesting how fruitful an analysis of the main legal language peculiarities can be. For this purpose, a comparative study of the specialised language used within legal documents with respect to the newswire language is carried out at a considerable level of detail. In particular, corpora of Italian and English legal texts have been parsed and compared with corpora of ordinary language in order to detect the main syntactic and lexical characteristics of legal language. The eventual goal is to suggest that a study phase of linguistic peculiarities of legal texts can have practical importance in many Legal Knowledge Management tasks.

2   Background and Motivation

Despite the urgent need for legal text semantic processing, according to McCarty [16], little attention has been paid within the AI&Law community to how NLP can contribute to Legal Knowledge Management. Rather, research in the field has been conducted mainly from a top–down perspective, uniquely stemming from domain–theoretical assumptions. So that, a bottom–up investigation of if and to what extent legal text semantic processing may benefit from linguistic analyses and techniques has been mostly overlooked.

Moreover, a corpus of law texts *linguistically annotated* (i.e. with morphological, syntactic and semantic information made explicit) is lacking, even though it would be useful for a number of Knowledge Management applications, such as Information Extraction and Domain Ontology Learning (as it has been recently raised in McCarty [17]).

However, in the last few years the number of NLP–oriented research activities has increased as witnessed by the workshops and tutorials recently organized on this topic. As a matter of fact, a survey of the main Knowledge Management applications in the legal domain can show that NLP tools are currently exploited in various studies. This is the case of the following cases:

1. **Legal Ontology Learning**, carried out by Van Gog et al. in [32], Lame in [11], Saias et al. in [28], Walter et al. in [34] and Völker et al. in [31];
2. **Legal Information Extraction**, carried out by Walter in [35] and McCarty in [16];
3. **Legal Semantic Annotation**, carried out by Bartolini et al. in [5], Brighi et al. in [6] and Spinosa et al. in [29];
4. **Automatic identification of legal terms** for lexicographic purposes, carried out by Pala et al. in [23] and [24];
5. **Legal Knowledge Modeling**, carried out by Nakamura et al. in [21];
6. **Legal Argumentation**, carried out by Moens et al. in [19], by Mochales et al. in [18] and by Wyner et al. in [36]; more recently, Wyner et al. in [37] put the focus on the need for bridging Computational Linguistics and Legal Argumentation efforts. In fact, the use of NLP tools is meant to support the formal construction of argument schemes;
7. **Legal Automatic Summarisation**, carried out by Grover et al. in [9].

However, in these studies and projects little attention has been paid:

1. to take into account the potentialities offered by each level of linguistic analysis (i.e. sentence splitter, tokenization in single word units, morphological analysis and shallow or deep syntactic parsing) to the following semantic processing of legal texts;
2. to put the emphasis on the need for domain–specific customizations asked for legal language peculiarities;
3. to point out legal language peculiarities with a special view to those characteristics which make this specialised language different with respect to ordinary language.

Interestingly enough, the vast majority of these works takes into account only the output of the component in charge of the *syntactic parsing* of the text. Mostly, the research activities considered here take into account the output of the *deep* level of syntactic parsing they have carried out. It follows that the previous levels of linguistic analysis, i.e. sentence splitter, tokenization in single word units and morphological analysis, are overlooked. However, according to what has been noted for the Biomedical Language Processing field in Ananiadou et al. [3], each level of linguistic analysis has been typically associated with specific processing components aimed at tackling domain–specific phenomena at some level. By focussing on the relationship between different processing components and various kinds of analyses, the authors allow appreciation of how each particular type of component relates to the overall Text Mining task.

In the legal domain, one exception is the analysis carried out by Pala and colleagues in [23], where results of the morphological analysis and lemmatizations of the Penal Code of the Czech Republic are presented. For this purpose, a morphological analyser designed for general Czech has been customised according to some legal language peculiarities, namely by adding legal terms. Interestingly, the authors put the focus on the main outcome of their work: as a result, they have obtained basic knowledge about the grammatical structure of legal texts (law terminology). Starting from the analysis of this processing component (i.e. the morphological analyser), they were further concerned in [24] with the development of a database containing valency frames of legal verbs, suitable for the description of the meanings of legally relevant verbs. In this respect, the Pala and colleagues' effort is aimed at exploring how even the morphological level of linguistic analysis can help the investigation of the semantic nature of legal language.

Secondly, just few of the works aforementioned are explicitly focused on the need for domain–specific customizations needed for legal language peculiarities. As witnessed by the efforts carried out in the Biomedical Language Processing community (see e.g. Lease et al. [12], Pyysalo et al. [25] and Sagae et al. [27]), studies overtly devoted to the adaptation of NLP tools for domain–specific purposes may improve the document processing task in terms of accuracy. An exception in the legal domain is represented by McCarty who in [16] developed a Definite Clause Grammar (DCG), consisting of approximately 700 rules, to result in "deep semantic interpretations" of a corpus of judicial opinions of the United States Supreme Court. He aimed at extracting the information that a lawyer wants to know about a case. He started from a qualitative analysis of general–purpose statistical parser (the Collins' parser) applied to those legal texts in order to test how accurate it was on sentences from judicial opinions. The parser results were mostly weak with respect to prepositional phrase attachments and coordinative conjunctions. Consequently, he foresaw several steps devoted to improving the accuracy of the parser for legal texts. It is also the case for Mochales and colleagues, who in [18] focussed on how legal language peculiarities are reflected in Argumentation Mining in legal texts. In order to detect and classify argumentative sentences in texts, the authors firstly looked for clauses of sentences on the basis of a predefined set of linguistic features, such e.g. typical verbs, significant sequences of two or more words, keywords, punctuation marks, etc. Then, using the linguistic characteristics of legal argument found in the previous phase, they defined a Context Free Grammar specifically devoted at parsing the argumentation structure in legal texts.

Finally, in the AI&Law field efforts devoted to investigating legal language peculiarities seem to be lacking. As will be demonstrated in the work presented here, such a study can help to shed light on those linguistic characteristics of legal documents, which might hamper Legal Knowledge Management efforts.

A significant exception is represented by the study by Nakamura and colleagues in [21], where the authors performed the linguistic investigation of a corpus of Japanese legal texts. Taking into account the linguistic characteristics they detected, they realised a system, which generates a logical formula corresponding to an input sentence from legal documents. In fact, they demonstrated that the results of this preliminary linguistic investigation are suitable for i) improving the accuracy of their NLP component that carries out the deep syntactic parsing of legal texts and ii) coping with particular legal sentence constructions that are difficult to transform into their corresponding logical forms. In particular, they put the focus on the analysis of typical Japanese nominalisations (i.e. noun phrases systematically related to the corresponding verbs), consisting of two nouns *A* and *B* with an adnominal particle "no", which carries some relation between *A* and *B*. The importance of an '*A no B*' relation relies on the fact that it is regarded as a verb. Frequently occurring in legal texts, these noun phrase types address the need for specific processing, in order to be transformed into a logical form, which expresses an event.

3 The Approach

This paper intends to continue the study carried out in Venturi [33] with a view to the practical importance that an investigation of the linguistic characteristics of legal texts has for Legal Knowledge Management purposes. In that previous study, the relative distribution of legal sub–language peculiarities has been identified by comparing the syntactic features detected in a corpus of Italian legal texts with the output of the syntactic parsing performed on a corpus of Italian ordinary language.

In the present study, a similar constrastive approach has been followed. Namely, syntactic and lexical characteristics of Italian and English legal language are identified by comparing an Italian and an English **legal** corpus with a reference corpus of Italian and English **ordinary** language. Afterwards, detected Italian and English legal language peculiarities are compared in order to investigate if, and to what extent, domain–specific characteristics are shared.

Syntactic and lexical levels of linguistic analysis have been carried out on the basis of the output of an NLP syntactic parsing component. In particular, the results presented here rely on a *shallow syntactic level of analysis*. As will be shown in Section 4, this paper maintains the widespread idea that a *shallow parsing* approach can help to provide enough detailed linguistic information for syntactically complex texts. Due to the minimal linguistic knowledge (i.e. morphosyntactic, lemma and word order information) a shallow syntactic component of analysis requires, such a level of analysis can be suitable to provide unambiguous syntactic representations.

4 NLP Analysis of Legal Texts

Syntactic and lexical levels of linguistic analysis are the focus of the present study. In particular, the latter level concerns *chunking*, the shallow syntactic parsing technique, which segments sentences into an unstructured sequence of syntactically organised texts units called *chunks*. Abney in [1] demonstrated how chunking proves to be a highly versatile means to produce reliable syntactic annotations of texts. The purpose of traditional full–parsing is to associate to each sentence a fully specified recursive structure, in order to identify the proper syntagmatic composition, as well as the relations of functional dependency among the identified constituents. On the contrary, chunking refers to a process of non–recursive segmentation of text. The resulting analysis is flat and unambiguous: only those relations which can be identified with certainty have been found out. Accordingly, some of the ambiguous grammatical dependencies (e.g. prepositional phrase attachments) are left underspecified and unresolved. This makes chunking highly suitable for the syntactic annotation of different types of texts, both written and spoken, and the analysis of corrupted or fragmentary linguistic inputs. According to Li et al. [14], as long as "parse incompleteness" is reinterpreted as "parse underspecification", failures due to lexical gaps, particularly complex syntactic constructions, etc. are minimised.

A number of reasons for carrying out a *shallow parsing* of legal texts are the following. According to Li et al. [14], in many natural language applications, such as Information Extraction and Text Summarisation, it is sufficient to use shallow parsing information, rather then relying on a deep syntactic analysis.

Although it might seem that full parsing should be preferred for adequate processing of texts, a shallow parsing approach has been chosen within some domain–specific applications. This is the case, for example, for Grover and colleagues, who in [9] investigated a method for generating flexible summaries of legal documents, by detecting a set of argumentative roles (e.g. fact, background, proceedings, etc.). Relying on the output of a chunking component of analysis, the authors carried out a *fact extraction* task from a corpus of judgments of the House of Lords.

Moreover, Bartolini et al. [5] and Spinosa et al. [29] have shown in their works the main advantages in taking chunked syntactic structure as the basis on which further stages of legal text processing operate. It has been reported there that chunked representations can profitably be used as the starting point for partial functional analyses, aimed at reconstructing the range of dependency relations within the law paragraph text that are instrumental for the semantic annotation of text. The major potential for text chunking lies in the fact that chunking does not "balk" at the domain–specific constructions that do not follow general grammar rules; rather it actually carries on parsing, while leaving behind any chunk unspecified for its category.

5   Parsing Italian Legal Texts

5.1   The NLP Tools

AnIta (Bartolini et al. [4]) is the parsing system used for the analysis of Italian legal texts. It is a general–purpose parsing system, which has already been tested as a component both in the SALEM semantic annotation system of legal texts

(Bartolini et al. [5]) and in the MELT (Metadata Extraction from Legal Texts) system (Spinosa et al. [29]) showing encouraging results. *AnIta* is constituted by a pipeline of NLP tools, which also includes a chunking module, CHUG–IT (Federici et al. [7]). In CHUG–IT chunking is carried out through a finite state automaton which takes as input a morpho-syntactically tagged text. According to Federici et al. [7], a *chunk* is a textual unit of adjacent word tokens; accordingly, discontinuous chunks are not allowed. Word tokens internal to a chunk share the property of being mutually linked through dependency relations which can be identified unambiguously with no recourse to lexical information other than part of speech and lemma. A sample output of this syntactic processing stage is given in Figure 1, where the input sentence is segmented into four chunks. Please note that each chunk contains information about its type (e.g. a noun chunk, N_C, a finite verb chunk, FV_C, a prepositional chunk, P_C, etc.), its lexical head (identified by the label POTGOV) and any occurring modifier and preposition.

Le stesse disposizioni si applicano ad un prodotto importato
'The same provisions are applied to an imported product'
```
[[CC:N_C][DET:LO#RD][PREMODIF:STESSO#A][POTGOV:DISPOSIZIONE#S]]
[[CC:FV_C][CLIT:SI#PQ][POTGOV:APPLICARE#V]]
[[CC:P_C][PREP:AD#E][DET:UN#RI][POTGOV:PRODOTTO#S]]
[[CC:ADJPART_C][POTGOV:IMPORTARE#V@IMPORTATO#A]]
```

Fig. 1. CHUG–IT output

The chunked sentence in Figure 1, shows an example of the use of under-specification. The chunking process resorts to underspecified analyses in cases of systematic ambiguity, such as the one between adjective and past participle. This ambiguity is captured by means of the underspecified chunk category ADJPART_C, subsuming both an adjectival chunk and a participal chunk interpretation.

This underspecified approach to robust syntactic analysis of Italian texts has been proved to be fairly reliable. Lenci et al. [13] provides a detailed evaluation of CHUG–IT parsing performance drawn on a corpus of financial newspapers articles. Results of automatic chunking were evaluated against a version of the same texts chunked by hand; they give a recall of 90.65% and a precision of 91.62%.

In what follows we wil provide an analysis of a corpus of Italian legal texts. For this purpose, the output of the chunking module included in *AnIta* (i.e. CHUG–IT) has been analyzed.

5.2 The Corpora

For the construction of the Italian legislative corpora two different design criteria were taken into account, namely the regulated domain and the enacting authority. The corpus is made up of legal documents which a) regulate two different domains, i.e. the environmental and the consumer protection domains and b) which are enacted by three different authorities, i.e. European Union, Italian state and Piedmont region.

The Environmental Corpus. The environmental corpus consists of 824 legislative, institutional and administrative acts for a total of 1,399,617 word tokens. It has been downloaded from the BGA (*Bollettino Giuridico Ambientale*), database edited by the Piedmont local authority for the environment[1]. The corpus includes acts enacted by the European Union, the Italian state and the Piedmont region, which cover a nine–year period (from 1997 to 2005). It is a heterogeneous document collection (henceforth referred to as Environmental Corpus) including legal acts such as national and regional laws, European directives, legislative decrees, as well as administrative acts, such as ministerial circulars and decision.

The Consumer Law Corpus. The corpus containing legal texts which regulate the consumer protection domain is a more homogeneous collection. Built and exploited in the DALOS project (Agnoloni et al. [2]), it is made up of 18 European Union Directives in consumer law (henceforth referred to as Consumer Law Corpus), for a total of 74,210 word tokens. Unlike the Environmental Corpus, it includes only Italian European legal texts.

5.3   Comparative Syntactic and Lexical Analysis

The investigation of syntactic and lexical peculiarities of legal language has been carried out starting from the chunked text (i.e. the output of CHUG–IT). The analysis mainly concerns:

1. the distribution of single chunk types;
2. the distribution of sequences of chunk types, with a view to those sequences which contain prepositional chunks;
3. the linguistic realization of events (i.e. situations) in legal texts.

A comparative method was followed. The distribution percentages of both single chunk types and sequences of chunks occurring within the Italian Legislative Corpus (i.e. the Environmental and the Consumer Law Corpus) were compared with the analysis of an Italian reference corpus, the PAROLE corpus (Marinelli et al., [15]), made up of about 3 million words including texts of different types (newspapers, books, etc.). Similarly, the typical linguistic realization of events in legal texts was highlighted by comparing the different lexical realization of situations depicted in legal documents and in the Italian reference corpus.

Distribution of Single Chunk Types. The distribution of single chunk types within legal texts was computed by comparing the occurrences of chunk types in the Italian Legislative Corpus and in the Italian reference corpus. This comparative approach is strengthened by the Chi-squared test applied on the obtained

---

[1] http://extranet.regione.piemonte.it/ambiente/bga/

results. It confirms the existence of a significant correlation between corpus variation and chunk type distribution.

Results of the parsing process, reported in Table 1, can help to highlight some main linguistic peculiarities of the Italian legal language and some consequences for Legal Knowledge Management. In particular, Table 1 shows the distribution of single chunk types in the Italian Legislative Corpus and in the Italian reference corpus. In this table, the count and the percentual frequency of occurrence are reported for each chunk type. It should be noted that the distribution of chunk types within the Environmental Corpus and the Consumer Law Corpus are kept distinct. As will be discussed in what follows, this choice of analysis brought about a number of related issues.

Table 1. Comparative distribution of single chunk types

Chunk types	Italian Legislative Corpus				PAROLE corpus	
	Environmental Corpus		Consumer Law Corpus			
	Count	%	Count	%	Count	%
Adj/Participial_C	38607	3.56	1689	2.74	29218	1.90
Adjectival_C	126267	11.66	6146	10.00	65740	4.27
Adverbial_C	13021	1.20	1006	1.63	49038	3.19
Coordinating_C	59585	5.50	3095	5.03	73073	4.75
Finite Verbal_C	36838	3.40	3007	4.89	140604	9.14
Nominal_C	226529	20.92	13062	21.25	413821	26.92
Non Finite Verbal_C	19569	1.80	5867	9.54	41674	2.71
Predicative_C	13047	1.20	843	1.37	21772	1.41
Prepositional_C	321167	29.66	14152	23.03	338037	21.99
Punctuation_C	192419	17.77	9756	15.87	278897	18.14
Subordinating_C	22026	2.03	2288	3.72	70226	4.56
Unknown_C	13439	1.24	535	0.87	14964	0.97

Interestingly enough, Table 1 shows that **prepositional chunks** (Prepositional_C) are the most frequent chunk types within the whole Italian Legislative Corpus. On the contrary, **nominal chunks** (Nominal_C) are the most recurring chunk types within the reference corpus. However, it should be appreciated that prepositional as well as nominal chunks are differently distributed between the Environmental Corpus and the Consumer Law Corpus. Namely, in the Environmental Corpus prepositional chunks constitute 29.66% of the considered chunks while the nominal chunks are 20.92%; in the Consumer Law Corpus the former ones are 23.03% while the latter ones are the 21.25%. Conversely, in the Italian reference corpus the nominal chunks are 26.92% of the total amount of chunk types and the prepositional chunks are 21.99%.

Moreover, a fairly low percentage of **finite verbal chunks** seems to be one of the main specific features of legal texts. Whereas the Italian reference corpus has 9.14% of the finite verbal chunks, their occurrence is about a third of that within the Environmental Corpus, i.e. 3.40%, and they only constitute 4.89% of the total amount of considered chunk types in the Consumer Law Corpus.

Various remarks follow from the results obtained by this first level of shallow parsing. First, the different distributions of single chunk types within the two analysed corpora of legal texts raised the need for a finer–grained investigation of legal corpora. Such a further analysis took into account that this difference might be due to the different enacting authorities, i.e. the Italian state and the Piedmont region, which enacted two–thirds of the Environmental Corpus, and the European Union, which enacted both one–third of the Environmental Corpus and the whole Consumer Law Corpus. In order to investigate this hypothesis, we investigated the distribution of single chunk types within the three sub–corpora, which made the Environmental Corpus.

Table 2. Comparative distribution of single chunk types within three Environmental sub–corpora

Chunk Types	Italian Legislative Corpus					
	Environmental Corpus					
	Region		State		Europe	
	Count	%	Count	%	Count	%
Adj/Participial_C	7247	3.58	20305	3.58	11055	3.52
Adjectival_C	24949	12.33	68931	12.16	32387	10.33
Adverbial_C	2149	1.06	5944	1.04	4928	1.57
Coordinating_C	10315	5.09	31930	5.63	17340	5.53
Finite Verbal_C	5857	2.89	16601	2.92	14380	4.58
Nominal_C	42850	21.17	114404	20.18	69275	22.10
Non Finite Verbal_C	3509	1.73	7927	1.39	8133	2.59
Predicative_C	1850	0.91	6467	1.14	4730	1.50
Prepositional_C	59615	29.46	175011	30.87	86541	27.61
Punctuation_C	36373	17.97	103696	18.29	52350	16.70
Subordinating_C	3348	1.65	10068	1.77	8610	2.74
Unknown_C	4279	2.11	5496	0.96	3664	1.16

Results of this investigation are reported in Table 2, where the count and the percentual frequency of occurrence are shown for each chunk type. By keeping distinct the three different enacting authorities, different syntactic peculiarities of the legal language used in the European Italian legal texts and in the national and local legal texts were highlighted. Interestingly, it seems that the Italian European legal language has linguistic features which make it more similar to ordinary language than the national and local legal language. Table 2 shows in particular that the Environmental sub–corpus made up by legal texts enacted by the Italian state is characterised by the highest occurrence of prepositional chunks; these are 30.87% of the total amount of considered chunk types. They show a slightly lower occurrence in the Environmental sub–corpus made up by legal texts enacted by the Piedmont region, where the prepositional chunks are 29.46%, and it is 27.61% in the European part of the Environmental

Corpus. Moreover, the distribution of finite verbal chunks in the three sub–corpora strengthened the first hypothesis. They are 2.89% and 2.92% respectively in the local and in the national sub–corpus; while they occur twice as much in the European sub–corpus, i.e. 4.58%. Interestingly, this latter percentage distribution of finite verbal chunks is more similar to the corresponding distribution of this chunk type within the Italian reference corpus (i.e. 9.14%).

Thus, this comparative analysis resulted in a close relationship between the European Italian legal texts and the Italian reference corpus, closer than the relationship between the latter and the national and local legal documents. It seems to suggest that the European legislator, to a certain extent, took into account the frequently advocated plain language recommendations. In other words, the language used during the legal drafting process of European legal documents reveals itself as less different from ordinary language. It follows that the processing of European legal language may require fewer customizations of NLP tools due to legal language peculiarities than the processing of national and local legal texts. Consequently, Legal Knowledge Management applications in the European field will be less hampered by linguistic obstacles caused by domain–specific features.

Moreover, the two more visible syntactic peculiarities, i.e. the higher occurrence of prepositional chunks and the lower presence of finite verbal chunks, detected within the whole Italian Legislative Corpus with respect to the Italian reference corpus, raised the need for exploring two hypotheses. The first concerns the possibility that such a high occurrence of prepositional chunks is strongly connected with their presence within sequences of chunks. As it will be described in the "Distribution of Sequences of Chunk Types" Section, according to this hypothesis, the distribution of sequences of certain chunk types has been investigated. A special focus has been put on those sequences which contain prepositional chunks. The second hypothesis concerns the bias typical of legal texts towards a *nominal* realization of events (situations) rather than a *verbal* realization. The observed low occurrence of finite verbal chunks gave rise to this hypothesis. Accordingly, in the "Linguistic Realization of Events in Legal Texts" Section, an investigation will be carried out into how events are more typically expressed within the Italian Legislative Corpus with respect to the Italian reference corpus.

Distribution of Sequences of Chunk Types. The hypothesis made about by the high occurence of prepositional chunks within the Italian Legislative Corpus concerned the possibility that these chunk types would be typically contained in sequences of chunks. In particular, a hypothesis was put forward regarding the presence of long sequences which include a high number of embedded prepositional chunks.

In order to test this hypothesis, sequences of chunk types containing prepositional chunks have been automatically identified. The following typology of cases has been considered:

1. chains of consecutive prepositional chunks, such as the following excerpt
   presentazione delle domande di contributo ai Comuni per l'attivazione dei

distributori per la vendita di metano ([N_C presentazione] [P_C delle do-
mande] [P_C di contributo] [P_C ai Comuni] [P_C per l'attivazione] [P_C di
distributori] [P_C per la vendita] [P_C di metano]) "submission of contribution requests to Municipalities for the activation of distributors for the sale
of natural gas";

2. sequences of prepositional chunks with possibly embedded adjectival chunks,
   such as the following excerpt *disciplina del canone regionale per l'uso di
   acqua pubblica* ([N_C disciplina] [P_C del canone] [ADJ_C regionale] [P_C
   per l'uso] [P_C di acqua] [ADJ_C pubblica]) "regulation of the regional fee
   for public water usage";

3. sequences of prepositional chunks with possibly embedded adjectival chunks,
   coordinative conjunctions and/or "light" punctuation marks (i.e. comma),
   such as the following excerpt *acqua destinata all'uso igienico e potabile,
   all'innaffiamento degli orti* ... ([N_C acqua] [ADJPART_C destinata] [P_C
   all'uso] [ADJ_C igienico] [COORD_C e] [ADJ_C potabile] [PUNC_C,] [P_C
   all'innaffiamento] [P_C degli orti]) "water devoted to sanitary and drinkable
   usage, to garden watering".

The investigation especially focused on the different distribution of deep chains
containing prepositional chunks (referred to as *PP–chains*) in the different kinds
of texts considered. Results are shown in Table 3, which shows the count of
embedded PP–attachments (i.e. sequences of chunk types containing embedded
prepositional chunks) that occurred within a sentence of legal texts with respect
to an ordinary language sentence [2].

By inspecting Table 3, the occurrence of deep PP–chains does not prove to
be a special syntactic feature of legal language with respect to ordinary language. Rather, the crucial distinguishing characteristic of the Italian Legislative
Corpus appears to be the *different percentual distributions of deeply embedding
sequences containing prepositional chunks*. Legal texts appear to have a higher
percentage of deep PP–chains with respect to the Italian reference corpus. Moreover, the analysis of different percentual occurrences within the three different
Environmental sub–corpora and within the Consumer Law Corpus allowed the
highlighting of finer–grained peculiarities. In general, it should be noticed that
there mainly are chains including 5 to 11 embedded chunks. For example, chains
of 8 PP–attachments constitute 5.78% of the total amount of PP-chains occurring within the legal texts enacted by the Piedmont region and 5.52% in
the documents enacted by the Italian state. Yet, theyhave a coverage of only
2.47% in the Italian reference corpus. As highlighted in the "Distribution of
Single Chunk Types" Section, the Italian European legal texts show a close
relationship with ordinary language. Accordingly, chains of 8 PP–attachments
have lower frequency of occurrence; they are 4.24% in the European part of the
Environmental Corpus and 3.26% in the Consumer Law Corpus.

These findings allow us to consider a number of statements. First, deep PP–
attachment chains seem to be typical of legal texts. They range from chains of

---

[2] Note that the first column of the Table above (named "PP–chains depth") reports the
number of chunk types embedded, with respect to the typology of cases considered.

Table 3. Comparative distribution of PP–attachment chains

PP–chains depth	Italian Legislative Corpus								PAROLE Corpus	
	Environmental Corpus						Consumer Law Corpus			
	Region		State		Europe					
	Count	%	Count	%	Count	%	Count	%	Count	%
4	2822	38.48	8924	37.42	4164	43.19	611	45.32	10240	54.72
5	1723	23.71	5366	22.50	2258	23.42	356	26.40	4621	24.68
6	1043	14.35	3505	14.69	1380	14.31	139	10.31	1999	10.68
7	612	8.42	2103	8.81	725	7.52	104	7.75	910	4.85
8	420	5.78	1318	5.57	409	4.24	44	3.26	464	2.47
9	248	3.41	813	3.40	237	2.45	28	2.07	206	1.09
10	151	2.13	652	2.73	161	1.67	23	1.70	112	0.59
11	91	1.35	350	1.46	92	0.95	10	0.74	74	0.39
12	63	0.88	244	1.02	69	0.71	7	0.51	39	0.20
13	30	0.42	167	0.70	39	0.40	9	0.66	28	0.14
14	19	0.32	147	0.61	37	0.38	5	0.37	17	0.09
15	18	0.28	79	0.33	27	0.28	1	0.07	6	0.03
16	11	0.25	62	0.25	26	0.27	6	0.44	5	0.02
17	6	0.09	40	0.16	5	0.05	1	0.07	3	0.01
18	3	0.05	31	0.12	4	0.04	3	0.22	2	0.01
19	3	0.04	24	0.10	3	0.03	0	0.00	1	0.00
20	2	0.02	23	0.09	4	0.04	1	0.07	3	0.00

embedded cross–reference to other legal documents, or sections of text (such as paragraphs, articles, etc.), such as the following sequence containing embedded prepositional chunks *all'articolo 1, comma 1, della legge 8 febbraio 2001, n. 12,* ... "in article 1, paragraph 1, of the act 8 February 2001, n. 12, ...", to chains of deverbal nouns (i.e. nouns morphologically derived from verbs), such as the following example *ai fini dell'accertamento della compatibilità paesaggistica* ... "to the verification of the landscape compatibility ...". In both cases, detecting these kinds of deep PP–chains would be fruitful for legal document transparency. As a matter of fact, the recurrence of complex and ambiguous syntactic constructions, such as deep sequences of prepositional chunks, is widely acknowledged to be responsible for the lack of understandability of legal texts. According to Mortara Garavelli [20], it is not the occurrence of abstract deverbal nouns which may affect the whole legal text comprehension; rather, the complex syntactic patterns, in which these deveba nouns are typically embedded, make legal texts difficult to comprehend. This is in line with some findings in studies on linguistic complexity, mainly in the cognitive and psycholinguistic field (see Fiorentino [8] for a survey of the state–of–the–art). It was discovered that our short term memory is able to receive, process and remember an average of 7 linguistic units. In processing a given input sentence the language user attempts to obtain closure on the linguistic units contained in it as early as possible. Thus, it is perceptually "costly" to carry on analysing deep chains of embedded sentence constituents.

Finally, as mentioned above, the analysis of sequences of prepositional chunks containing **deverbal nouns** may be related to a study of the linguistic realization of events (situations) in legal texts. Let us consider the two following sentences:

1. l'autorità amministrativa competente accerta la compatibilità paesaggistica ("the relevant administrative authority verifies the landscape compatibility"),
2. il Comune è preposto alla gestione del vincolo ai fini dell'accertamento della compatibilità paesaggistica ... ("the Municipality is in charge of the management of the obligation to the verification of the landscape compatibility").

In the first case, the event "verification" is realised through a verbal construction involving the verb *accertare* ('to verify'). In the second sentence, the same event is realised through a nominal construction headed by the deverbal noun *accertamento* ('verification'). Interestingly, it should be noted that in the latter case the deverbal noun is embedded in a sequence of prepositional chunks, i.e. ... *preposto alla gestione del vincolo ai fini dell'accertamento della compatibilità paesaggistica* ... "... in charge of the management of the obligation to the verification of the landscape compatibility ...". According to these findings, remarks on Legal Knowledge Management applications such as Event Extraction can benefit from the results obtained by an analysis of PP–chains containing deverbal nouns.

Linguistic Realization of Events in Legal Texts. The low percentual occurrence of finitive verbal chunks found in Section 5.3 hinted at lexical realization patterns of situations and events, which is typical of legal documents. In order to follow this direction of research, a case study was carried out on a small sample of some main events within the Italian Legislative Corpus and the Italian reference corpus.

The results are reported in Table 4, where for each event type the corresponding verbal and nominal morpho–syntactic realization is shown in the second column. It should be noted that the percentual occurrence of the type of morpho–syntactic realization has been computed as the ratio of the noun (or of the verb) occurrence over all types of realization (i.e. nominal + verbal) of a given event. In the last columns of the table, the count and the percentual occurrence of the two linguistic realization types are shown. Interestingly, it highlights a broad bias towards a **nominal** realization of same main events within the Italian Legislative Corpus.

This is the case for the 'Violate' event triggered by words which convey a situation where someone or something violates a set of rules. As shown in Table 4, this event type can be expressed by the verb 'violare' (to violate) and by the deverbal noun 'violazione' (infringement). **Nominal** realization was more frequent in the Italian Legislative Corpus than in the Italian reference corpus. However, a different percentual occurrence can be seen in the legal texts enacted by the European Union (both regulating the environmental and consumer protection domain), and in the documents enacted by the local authority and by the Italian state. According to previous findings, the local and national legal

Table 4. Comparative morpho–syntactic realization of events

Event type	Morpho-syntactic realization	Italian Legislative Corpus				PAROLE Corpus	
		European texts		Regional & national texts			
		Count	%	Count	%	Count	%
ENFORCE	attuare (to enforce)	159	24.02	184	9.94	88	43.35
	attuazione (enforce-ment)	503	75.98	1668	90.06	115	56.65
VIOLATE	violare (to violate)	8	9.09	5	2.94	107	52.97
	violazione (infringe-ment)	80	90.91	165	97.06	95	47.03
PROTECT	proteggere (to pro-tect)	107	16.61	296	26.35	179	55.59
	protezione (protec-tion)	537	83.39	819	73.45	143	44.41
IMPOSING_OBLIGATION	obbligare (to obli-gate)	19	6.01	59	8.18	122	42.21
	obbligo (obligation)	297	93.99	662	91.82	167	57.79

texts seem to contain more domain–specific peculiarities. In those documents, the 'Violate' event is realized in 97.06% of the total amount of cases by the deverbal noun 'violazione' (infringement) and only in 2.94% of cases by the verb 'violare' (to violate). In the European documents there is also a strong bias towards the nominal realization, however with different occurrence percentages: the deverbal noun 'violazione' (infringement) occurs in 90.91% of all 'Violate' event realizations, and the verb 'violare' (to violate) is 9.09% of cases.

Conversely, within the Italian reference corpus the variance of morpho–syntactic realization type shows different characteristics. Not only is the verbal realization more frequent than the nominal one – 52.97% versus 47.03% respectively –, but also, it seems that ordinary language does not have any sharp bias towards one of the two types of linguistic realization.

These findings prompted an assessment of the consequences for Legal Knowledge Management tasks such as Event Extraction from legal document collections. According to the state–of–the–art literature in the Event Knowledge Management field, the processing of nouns and deverbal nouns is as crucial as challenging. As it is claimed in Gurevich et al. [10], deverbal nouns, or *nominalizations*, pose serious challenges for general–purpose knowledge–representation systems. They report that the most common strategy to face with this relevant problem involves finding ways to create verb–like representations from sentences which contain deverbal nouns, i.e. strategies to map the arguments of deverbal nouns to those of the corresponding verbs. In particular, tasks such as Semantic Role Labelling for event nominalizations [22] are very concerned with this challenge.

The results shown in this section reveal that, more than within the open–domain field, the Event Knowledge Management task in the legal domain is made more challenging by the rather high occurrence of nouns and deverbal nouns, which should be considered *event predicative.*

6   Parsing English Legal Texts

The syntactic and lexical analysis of English Legal texts adopted the same criteria applied to the investigation of Italian legal texts. Accordingly, it relies on a shallow level of syntactic parsing (i.e. chunking), and it is carried out by following a similar comparative method. As in the case of the linguistic analysis of Italian legal language, the relative distribution of English legal language characteristics has been investigated with respect to an English reference corpus. As it will be described in Section 7, comparing the results obtained by parsing Italian and English legal texts, the eventual goal is to investigate whether some syntactic and lexical peculiarities were shared by the Italian and English legal language.

6.1   The NLP Tools

The *GENIATagger* (Tsuruoka et al. [30]), a NLP component carrying out part–of–speech tagging and chunking, has been used to perform the English legal texts analysis. Even though the output of this component is quite similar to CHUG–IT's, the output of the two tools differs to some extent. In fact, they mainly diverge because of different grammatical requirements in the two languages considered (i.e. Italian and English), as well as differences in linguistic annotation choices.

The fragment of *GENIATagger* chunked text, reported in Table 5, shows how the *GENIATagger* outputs. In the first column (Word Form) the word is shown as it appears in the original sentence; the second column lists the lemma of the word; the part–of–speech tag is in the third column (e.g. NN is the tag used for nouns, IN is the tag which labels prepositions other than *to*, etc.). The last column indicates the chunk type (e.g. NP indicated a nominal chunk, PP is a prepositional chunk, etc.). It should be noted that chunks are represented in the IOB2 format; thus, in the Table B stands for BEGIN (of the chunk) and I for INSIDE (the chunk itself).

In particular, it should be noticed that the output of the *GENIATagger* and that of CHUG–IT mostly differ because of their representation of nominal and prepositional chunks. A prepositional chunk does not contain anything more than a preposition inside, such as *to* or *as* which are at the beginning of the PP chunk (i.e. B–PP). This annotation strategy is relevant for the English syntactic features concerning the stranding of prepositions within a sentence. Conversely, a nominal chunk can be a textual unit of adjacent word tokens, such as *certain exonerating circumstances*, which includes an adjective (*certain*, JJ) at the beginning (B–NP), an introducing present participle (*exonerating*, VBG) and a common noun (*circumstances*, NNS) as two inner elements (I–NP). Yet, it can also be made up of a single word token, such as *proof*, which includes a common noun (NN) only, or *he*, which is made up of a personal pronoun (PRP). However, it can never include a preposition.

On the contrary, the annotation strategy of CHUG–IT allows segmenting a sentence differently. As reported in Section 5.1, for example, the prepositional chunk *ad un prodotto*, "to a product", always includes both the preposition *a* ("to") and the noun *prodotto* ("product").

Table 5. GENIATagger annotation

Word Form	Lemma	Part-Of-Speech	Chunk Type
he	he	PRP	B-NP
proof	proof	NN	B-NP
furnishes	furnishes	VBZ	B-VP
as	as	IN	B-PP
to	to	TO	B-PP
the	the	DT	B-NP
existence	existence	NN	I-NP
of	of	IN	B-PP
certain	certain	JJ	B-NP
exonerating	exonerating	VBG	I-NP
circumstances	circumstances	NNS	I-NP

"...he furnishes proof as to the existence of certain exonerating circumstances ..."

6.2 The Corpus

For the English legal text analysis, a collection of 18 English European Union Directives in consumer law has been used. The corpus has been built and exploited in the DALOS project (Agnoloni et al. [2]). It is made up of the English version of the Italian corpus in consumer law. This legal corpus has been compared with a sub–corpus of the Wall Street Journal made up of 1,138,189 words, which was used as a reference corpus.

6.3 The Comparative Syntactic and Lexical Analysis

Differently from the Italian case, the comparison between the English Legislative Corpus and the reference corpus (i.e. WSJ Corpus) has concentrated on:

1. the distribution of single chunk types,
2. the linguistic realization of events (i.e. situations) in legal texts.

A more exhaustive syntactic investigation is still ongoing, also including the analysis which concerns the distribution of sequences of chunk types, compared to those sequences which contain prepositional chunks.

Distribution of Single Chunk Types. The distribution comparison of chunk types, between the English Legislative Corpus in consumer law and the reference corpus, shows some legal language peculiarities wich have been detected previously for Italian. As in the Italian legal texts, within the English legal documents the occurrence of **prepositional chunks** has been noted to be higher than in the general language texts (see Table 6). They constitute 27.21% of the total number of chunk types in the English European Union Directives, against 19.88% in the Wall Street Journal sub–corpus. At the same time, the percentage of **nominal chunks** is lower in legal texts (48.16%) than in the reference corpus, where they represent 51.84% of the identified chunks.

Regarding the distribution of **finite verbal chunks**, the comparative analysis shows that they have a quite low percentage of occurrence. In particular, they represent 9.17% of the total chunk types within the English legislative corpus, compared to 15.56% in the reference corpus.

Table 6. Comparative distribution of chunk types

Chunk Types	English Legislative Corpus		WSJ Corpus	
	Count	%	Count	%
Nominal_C	17731	48.16	336635	51.84
Prepositional_C	10019	27.21	129131	19.88
Finite verbal_C	3378	9.17	101092	15.56
Non finite verbal_C	2401	6.52	26673	4.10
Adverbial_C	835	2.26	24139	3.71
Adjectival_C	823	2.23	11726	1.80

It should be noted that these results are in line with the ones from the analysis of the corpus of Italian legal texts enacted by the European Community. In fact, one of the most prominent findings in Section 5.3 was the close relationship between European Italian legal language and ordinary language. In that case, it was shown that such relationship is closer than the one between the legal language used in national and local documents, and ordinary Italian. However, the rather low frequency of finite verbal chunks found in European English legal texts suggest that these documents are possibly characterised by a significant bias towards a nominal realization of events.

Linguistic Realization of Events in Legal Texts. In order to investigate the hypothesis motivated by the low occurrence of finite verbal chunks in the English Legislative Corpus, a case study was carried out on a small sample of some main events. Similar to the Italian case study, the different lexical realizations of situations depicted in English legal texts and in the English reference corpus were investigated. The percentual occurrence of each type of morpho–syntactic realization was computed as the ratio of the noun (or of the verb) occurrence over all types of realization (i.e. nominal + verbal) of a given event. The results reported in Table 7 verify the first hypothesis, i.e. a broad bias typical of English legal documents towards the **nominal** realization of events.

In fact, in most of the cases reported in Table 7, the event nominal realizations are percentually more frequent in the legislative corpus than the verbal constructions; while an opposite bias has been observed within the WSJ reference corpus. Interestingly, as previously observed for the Italian case, the same 'Violate' event within the English Legislative Corpus is realized in 2.67% of the total amount of occurrences through the verb 'to violate', while the nominal realization through the noun 'infringement' is 97.33% of cases. On the contrary, within the WSJ Corpus the event occurs more frequently through a verbal construction, i.e. 86.59% of cases, than through a nominal one, i.e. 13.41%.

According to these findings, as has previously been discussed in see Section 5.3, the investigation of the linguistic realization of events in legal texts might be of great importance for Event Knowledge Management.

Table 7. Comparative morpho–syntactic realization of events

Event type	Morpho-syntactic realization	English Legislative Corpus		WSJ Corpus	
		Count	%	Count	%
ENFORCE	to enforce	8	14.81	17	26.56
	enforcement	46	85.19	47	73.44
VIOLATE	to violate	2	2.67	71	86.59
	infringement	73	97.33	11	13.41
PROTECT	to protect	64	27.35	116	45.14
	protection	170	72.65	141	54.86
PROHIBIT	to prohibit	10	23.26	40	80.00
	prohibition	33	76.74	10	20.00

7 Comparing Italian and English Legal Language Peculiarities

In order to investigate which syntactic and lexical peculiarities are shared by the Italian and English legal language, we compare the results obtained in Section 5 and Section 6 with respect to:

1. the distribution of single chunk types,
2. the linguistic realization of events (i.e. situations) in legal texts.

This multilingual comparison takes into account the results obtained by contrasting the sub–part of the Environmental Corpus made up of texts enacted by the European Union and the Italian Consumer Law Corpus with the English Legislative Corpus. This was made possible by the homogeneous nature of the three corpora: they are all enacted by the same enacting authority, i.e. the European Union. It follows that this comparison concerns those syntactic and lexical peculiarities shared by the European legal language used in the two considered corpora.

Despite the different grammatical requirements in the two languages considered (i.e. Italian and English) and the different annotation choices of the two NLP tools exploited (i.e. the *GENIATagger* and CHUG–IT), the distribution of single chunk types between Italian and English European legal texts shows similarities. Comparing the two European legal languages, some main features have been revealed as shared, namely:

1. a high occurrence of prepositional chunks,
2. a fairly low presence of finite verbal chunks.

The percentual distribution of these two chunk types within the considered legal corpora with respect to the corresponding distribution within the analysed reference corpora is significantly similar in the Italian and English cases. Namely:

1. within the European sub–part of the Italian Environmental Corpus, the prepositional chunks represent 27.61% of the total amount of chunks and 23.03% in the Italian Consumer Law Corpus; on the contrary, this chunk type coverss 21.99% in the Italian reference corpus;
2. within the European sub–part of the Italian Environmental Corpus, the finite verbal chunks cover 4.58% of the total amount of chunks and 4.89% in the Italian Consumer Law Corpus; on the contrary, this chunk type counts for 9.14% in the Italian reference corpus;
3. within the English Legislative Corpus, the prepositional chunks represent 27.21% of the total amount of chunks computed; on the contrary, this chunk type constitutes 19.88% in the English reference corpus;
4. within the English Legislative Corpus, the finite verbal chunks cover 9.17% of the total amount of chunks; on the contrary, this chunk type covers 15.56% in the English reference corpus.

Interestingly enough, it has been shown that both in the Italian case (see Section 5.3) and in the English case (see Section 6.3) the fairly low presence of finite verbal chunks within legal texts is closely related to a typical linguistic realization of events. However roughly, this shallow level of syntactic analysis shows a shared broad bias towards a *nominal* realization of some main events within Italian and English European legal texts.

8   Conclusion and Future Directions of Research

The results of an analysis of the main syntactic features of legal language detected within legal corpora have been presented in this article. Such an investigation relies on the output of an NLP component of analysis, which syntactically parses document collections at a *shallow level*. This output of *chunking components* has been analysed, and a three–level comparison has ben performed:

1. specialised and ordinary language,
2. different legal languages used by different enacting authorities (i.e. European Union, Italian state and Piedmont region),
3. two different European legal languages (i.e. Italian and English), assuming a multilingual perspective.

Even if quite rudimentary, this first level of syntactic grouping has allowed us to detect some main characteristics of legal language.

Among others, the quite high occurrence of prepositional chunks and the fairly low presence of finite verbal chunks have been considered as two of the more visible syntactic phenomena particular to legal sub–language. Interestingly, these main syntactic peculiarities observed are shared by the two languages considered (i.e. Italian and English).

The investigation of the reason why finite verbal chunks occur with a low frequency within legal texts led us to detect a broad bias within both Italian and English legal corpora towards a **nominal** realization of events rather than verbal. In the article, it has been pointed out how the outcome of such a linguistic study can have practical importance for Legal Knowledge Management tasks, such as Event Knowledge Management. According to the state–of–the–art literature, the nominal realization of events poses serious challenges for knowledge–representation systems. Consequently, the rather high occurrence of nominal realizations within legal corpora might cause difficulties for Event Knowledge Management in the legal domain. This is in line with the work carried out by Nakamura and colleagues in [21]. They started from the results obtained during a phase of investigation of the linguistic realization of events within Japanese legal texts. In particular, they put the focus on the analysis of typical Japanese nominalisations (i.e. noun phrases systematically related to the corresponding verbs) that frequently occur within legal texts. Consequently, the authors relied on a specific processing of these noun phrase types in order to transform them into a logical form which expresses an event.

According to the results described in the "Distribution of Sequences of Chunk Types" Section, the high occurrence of prepositional chunks found in the Italian Legislative Corpus appears to be related to the bias towards nominal realization of events within legal texts. In particular, according to the typology of chains of prepositional chunks, deep sequences of prepositional chunks containing deverbal nouns were found to be connected with the nominal realization of events. It has been pointed out how these findings can have practical importance for Event Knowledge Management.

Moreover, the high frequency of deep sequences of prepositional chunks has been related with the lack of understability of legal texts. These outcomes suggest how legal texts are difficult to be understood not only by human beings, but also by computational tools. In particular, deep PP–attachment chains can pose serious challenges for an NLP syntactic component in charge of a *dependency level* of syntactic analysis of Italian legal texts. In particular, the different syntactic aspects of legal corpora analysed when compared to the Italian reference corpus dramatically suggests the need for a training phase of NLP tools for deep parsing purposes.

Acknowledgements

This work is the result of research activities of the Dylan Lab (Dynamics of Language Laboratory) of the Institute of Computational Lnguistics (ILC–CNR) of Pisa and of the Department of Linguistics (Computational Linguistics division) of the University of Pisa. *AnIta* has been developed and provided by its members.

References

1. Abney, S.P.: Parsing by chunks. In: Berwick, R.C., et al. (eds.) Principled–Based Parsing: Computation and Psycholinguistics, pp. 257–278. Kluwer Academic Publishers, Dordrecht (1991)
2. Agnoloni, T., Bacci, L., Francesconi, E., Peters, W., Montemagni, S., Venturi, G.: A two–level knowledge approach to support multilingual legislative drafting. In: Breuker, J., Casanovas, P., Francesconi, E., Klein, M. (eds.) Law, Ontologies and the Semantic Web. IOS Press, Amsterdam (2009)
3. Ananiadou, S., McNaught, J. (eds.): Text Mining for Biology and Biomedicine. Artech House, London (2006)
4. Bartolini, R., Lenci, A., Montemagni, S., Pirrelli, V.: Hybrid Constraints for Robust Parsing: First Experiments and Evaluation. In: Proceedings of LREC 2004, Fourth International Conference on Language Resources and Evaluation, Centro Cultural de Belem, Lisbon, Portugal, May 26-28, pp. 795–798 (2004)
5. Bartolini, R., Lenci, A., Montemagni, S., Pirrelli, V., Soria, C.: Automatic classification and analysis of provisions in legal texts: a case study. In: Meersman, R., Tari, Z., Corsaro, A. (eds.) OTM-WS 2004. LNCS, vol. 3292, pp. 593–604. Springer, Heidelberg (2004)
6. Brighi, R., Lesmo, L., Mazzei, A., Palmirani, M., Radicioni, D.P.: Towards Semantic Interpretation of Legal Modifications through Deep Syntactic Analysis. In: Proceedings of the 21th International Conference on Legal Knowledge and Information Systems (JURIX 2008), Florence, December 10-13 (2008)
7. Federici, S., Montemagni, S., Pirrelli, V.: Shallow Parsing and Text Chunking: a View on Underspecification in Syntax. In: Carroll, J. (ed.) Proceedings of the Workshop on Robust Parsing, ESSLLI, Prague, August 12-16 (1996)
8. Fiorentino, G.: Web usability e semplificazione linguistica. In: Venier, F. (ed.) Rete Pubblica. Il dialogo tra Pubblica Amministrazione e cittadino: linguaggi e architettura dellinformazione, Perugia, Edizione Guerra, pp. 11–38 (2007)
9. Grover, C., Hachey, B., Hughson, I., Korycinski, C.: Automatic Summarisation of Legal Documents. In: Proceedings of the 9th International Conference on Artificial Intelligence and Law (ICAIL 2003), Scotland, United Kingdom, pp. 243–251 (2003)
10. Gurevich, O., Crouch, R.: Deverbal Nouns in Knowledge Representation. Journal of Logic and Computation 18(3), 385–404 (2008)
11. Lame, G.: Using NLP techniques to identify legal ontology components: concepts and relations. In: Benjamins, V.R., Casanovas, P., Breuker, J., Gangemi, A. (eds.) Law and the Semantic Web. LNCS (LNAI), vol. 3369, pp. 169–184. Springer, Heidelberg (2005)
12. Lease, M., Charniak, E.: Parsing Biomedical Literature. In: Dale, R., Wong, K.-F., Su, J., Kwong, O.Y. (eds.) IJCNLP 2005. LNCS (LNAI), vol. 3651, pp. 58–69. Springer, Heidelberg (2005)
13. Lenci, A., Montemagni, S., Pirrelli, V.: CHUNK–IT. An Italian Shallow Parser for Robust Syntactic Annotation. In: Linguistica Computazionale, Istituti Editoriali e Poligrafici Internazionali, Pisa–Roma (2001)

14. Li, X., Roth, D.: Exploring Evidence for Shallow Parsing. In: Proceedings of the Annual Conference on Computational Natural Language Learning, Toulouse, France (2001)
15. Marinelli, R., Biagini, L., Bindi, R., Goggi, S., Monachini, M., Orsolini, P., Picchi, E., Rossi, S., Calzolari, N., Zampolli, A.: The Italian PAROLE corpus: an overview. In: Zampolli, A., Calzolari, N., Cignoni, L. (eds.) Computational Linguistics in Pisa - Linguistica Computazionale a Pisa. Linguistica Computazionale, Special Issue, XVI-XVII, Pisa–Roma, IEPI. Tomo I, pp. 401–421 (2003)
16. McCarty, L.T.: Deep Semantic Interpretations of Legal Texts. In: Proceedings of the 11th international conference on Artificial intelligence and law (ICAIL), June 4-8. Stanford Law School, Stanford (2007)
17. McCarty, T.: Remarks on Legal Text Processing – Parsing, Semantics and Information Extraction. In: Proceedings of the Worskhop on Natural Language Engineering of Legal Argumentation, Barcelona, Spain, June 11 (2009)
18. Mochales Palau, R., Moens, M.-F.: Argumentation Mining: the Detection, Classification and Structure of Arguments in Text. In: Proceedings of the 12th international conference on Artificial intelligence and law (ICAIL), June 8-12. Universitat Autonoma de Barcelona, Barcelona (2009)
19. Moens, M.-F., Mochales Palau, R., Boiy, E., Reed, C.: Automatic detection of Arguments in Legal Texts. In: Proceedings of the 11th International Conference on Artificial Intelligence and law (ICAIL), June 4-8. Stanford Law School, Stanford (2007)
20. Mortara Garavelli, B.: Le parole e la giustizia. In: Divagazioni grammaticali e retoriche su testi giuridici italiani, Torino, Einaudi (2001)
21. Nakamura, M., Nobuoka, S., Shimazu, A.: Towards Translation of Legal Sentences into Logical Forms. In: Satoh, K., Inokuchi, A., Nagao, K., Kawamura, T. (eds.) JSAI 2007. LNCS (LNAI), vol. 4914, pp. 349–362. Springer, Heidelberg (2008)
22. Pado, S., Pennacchiotti, M., Sporleder, C.: Semantic role assignment for event nominalisations by leveraging verbal data. In: Proceedings of Coling 2008, Manchester, UK, August 18-22 (2008)
23. Pala, K., Rychlý, P., Šmerk, P.: Automatic Identification of Legal Terms in Czech Legal Texts. In: Francesconi, E., Montemagni, S., Peters, W., Tiscornia, D. (eds.) Semantic Processing of Legal Texts. LNCS (LNAI), vol. 6036, pp. 83–94. Springer, Heidelberg (2010)
24. Pala, K., Rychlý, P., Šmerk, P.: Morphological Analysis of Law texts. In: Sojka, P.H., Aleš-Sojka, P. (eds.) First Workshop on Recent Advances in Slavonic Natural Languages Processing (RASLAN 2007), pp. 21–26. Masaryk University, Brno (2007)
25. Pyysalo, S., Salakoski, T., Aubin, S., Nazarenko, A.: Lexical adaptation of link grammar to the biomedical sublanguage: a comparative evaluation of three approaches. BMC Bioinformatics 7(Suppl. 3), S2 (2006)
26. Rissland, E.L.: Ai and legal reasoning. In: Proceedings of the International Joint Conference in Artificial Intelligence, IJCAI 1985 (1985)
27. Sagae, K., Tsujii, J.: Dependency parsing and domain adaptation with LR models and parser ensembles. In: Proceedings of EMNLP–CoNLL 2007, pp. 1044–1050 (2007)
28. Saias, J., Quaresma, P.: A Methodology to Create Legal Ontologies in a Logic Programming Based Web Information Retrieval System. In: Benjamins, V.R., Casanovas, P., Breuker, J., Gangemi, A. (eds.) Law and the Semantic Web. LNCS (LNAI), vol. 3369, pp. 185–200. Springer, Heidelberg (2005)

29. Spinosa, P., Giardiello, G., Cherubini, M., Marchi, S., Venturi, G., Montemagni, S.: NLP–based Metadata Extraction for Legal Text Consolidation. In: Proceedings of the 12th International Conference on Artificial Intelligence and Law (ICAIL 2009), Barcelona, Spain, June 8-12 (2009)

30. Tsuruoka, Y., Tateishi, Y., Kim, J.-D., Ohta, T., McNaught, J., Ananiadou, S., Tsujii, J.: Developing a Robust Part–of–Speech Tagger for Biomedical Text. In: Bozanis, P., Houstis, E.N. (eds.) PCI 2005. LNCS, vol. 3746, pp. 382–392. Springer, Heidelberg (2005)

31. Völker, J., Langa, S.F., Sure, Y.: Supporting the Construction of Spanish Legal Ontologies with Text2Onto. In: Casanovas, P., Sartor, G., Casellas, N., Rubino, R. (eds.) Computable Models of the Law. LNCS (LNAI), vol. 4884, pp. 105–112. Springer, Heidelberg (2008)

32. Van Gog, R., Van Engers, T.M.: Modelling Legislation Using Natural Language Processing. In: Proceedings of the 2001 IEEE International Conference on Systems, Man, and Cybernetics, SMC 2001 (2001)

33. Venturi, G.: Linguistic analysis of a corpus of Italian legal texts. A NLP–based approach. In: Karasimos, et al. (eds.) Proceedings of First Patras International Conference of Graduate Students in Linguistics (PICGL 2008), pp. 139–149. University of Patras Press (2008)

34. Walter, S., Pinkal, M.: Automatic extraction of definitions from german court decisions. In: Proceedings of the COLING 2006 Workshop on Information Extraction Beyond The Document, Sydney, July 2006, pp. 20–28 (2006)

35. Walter, S.: Linguistic Description and Automatic Extraction of Definitions from German Court Decisions. In: Proceedings of the Sixth International Language Resources and Evaluation (LREC 2008), Marrakech, Morocco, May 28-30 (2008)

36. Wyner, A., Mochales-Palau, R., Moens, M.F., Milward, D.: Approaches to Text Mining Arguments from Legal Cases. In: Francesconi, E., Montemagni, S., Peters, W., Tiscornia, D. (eds.) Semantic Processing of Legal Texts. LNCS (LNAI), vol. 6036, pp. 60–79. Springer, Heidelberg (2010)

37. Wyner, A., van Engers, T.: From Argument in Natural Language to Formalised Argumentation: Components, Prospects and Problems. In: Proceedings of the Worskhop on Natural Language Engineering of Legal Argumentation, Barcelona, Spain, June 11 (2009)

Named Entity Recognition and Resolution in Legal Text

Christopher Dozier, Ravikumar Kondadadi, Marc Light,
Arun Vachher, Sriharsha Veeramachaneni, and Ramdev Wudali

Thomson Reuters Research and Development
Eagan, MN 55123, USA
{chris.dozier,ravikumar.kondadadi,marc.light,arun.vachher,
harsha.veeramachaneni,ramdev.wudali}@thomsonreuters.com
http://www.thomsonreuters.com

Abstract. Named entities in text are persons, places, companies, etc. that are explicitly mentioned in text using proper nouns. The process of finding named entities in a text and classifying them to a semantic type, is called named entity recognition. Resolution of named entities is the process of linking a mention of a name in text to a pre-existing database entry. This grounds the mention in something analogous to a real world entity. For example, a mention of a judge named *Mary Smith* might be resolved to a database entry for a specific judge of a specific district of a specific state. This recognition and resolution of named entities can be leveraged in a number of ways including providing hypertext links to information stored about a particular judge: their education, who appointed them, their other case opinions, etc.

This paper discusses named entity recognition and resolution in legal documents such as US case law, depositions, and pleadings and other trial documents. The types of entities include judges, attorneys, companies, jurisdictions, and courts.

We outline three methods for named entity recognition, lookup, context rules, and statistical models. We then describe an actual system for finding named entities in legal text and evaluate its accuracy. Similarly, for resolution, we discuss our blocking techniques, our resolution features, and the supervised and semi-supervised machine learning techniques we employ for the final matching.

Keywords: Named Entity Recognition, Named Entity Resolution, Natural Language Processing.

1 Introduction

Names are proper nouns and, in English, are usually capitalized and have other syntactic characteristics that differentiate them from other types of nouns. For example, names are often used without a determiner: *Thomas wrote in his dissent that the majority argument ...* Semantically, names can be thought to refer to a single entity in the world. Names play a central role in the information content of

E. Francesconi et al. (Eds.): Semantic Processing of Legal Texts, LNAI 6036, pp. 27–43, 2010.
© Springer-Verlag Berlin Heidelberg 2010

many types of human language, including legal texts: names are used to identify parties, attorneys, courts, jurisdictions, statues, judges, etc. involved in a legal proceedings. Thus, being able to recognize which word sequences are names and resolve these names to what they refer is useful for many legal text processing tasks. In this paper we describe methods for named entity (NE) recognition and resolution in legal texts.

To make this discussion more concrete, consider the text in Figure 1.

SUPREME COURT OF THE UNITED STATES
Syllabus
MICROSOFT CORP. v. AT&T CORP.
CERTIORARI TO THE UNITED STATES COURT OF APPEALS FOR THE FEDERAL CIRCUIT
No. 051056. Argued February 21, 2007—Decided April 30, 2007
It is the general rule under **United States** patent law that no infringement occurs when a patented product is made and sold in another country. There is an exception. Section 271(f) of the **Patent Act**, adopted in 1984, provides that infringement does occur when one "suppl[ies] . . . from the United States," for "combination" abroad, a patented invention's "components." 35 U. S. C. §271(f)(1). This case concerns the applicability of §271(f) to computer software first sent from the **United States** to a foreign manufacturer on a master disk, or by electronic transmission, then copied by the foreign recipient for installation on computers made and sold abroad. **AT&T** holds a patent on a computer used to digitally encode and compress recorded speech. **Microsoft**'s **Windows** operating system has the potential to infringe that patent because **Windows** incorporates software code that, when installed, enables a computer to process speech in the manner claimed by the patent. **Microsoft** sells **Windows** to foreign manufacturers who install the software onto the computers they sell. **Microsoft** sends each manufacturer a master version of **Windows**, either on a disk or via encrypted electronic transmission, which the manufacturer uses to generate copies. Those copies, not the master version sent by **Microsoft**, are installed on the foreign manufacturer's computers. The foreign-made computers are then sold to users abroad.

Fig. 1. Example legal text

Names are used to refer to the companies Microsoft and AT&T, the product Windows, the location United States, the court Supreme Court of the United States, and the section of U.S. Code called The Patent Act. Resolving such entities to particular entries in lists may be straightforward as in the case of the *Supreme Court of the United States* or might be more involved as in the *AT&T* because of the numerous companies related to AT&T (e.g., AT&T Wireless). *The Patent Act* is also ambiguous: there are many different versions of the Patent Act.

Humans are able to spot such names and disambiguate their references; this ability is part of understanding a legal text. For machines, recognizing and resolving names is an initial step towards making the semantics of the legal text explicit and available for further processing. This might include placing ID numbers resulting from resolution in search engine indexes. Users can then search for specific entities instead of words. The user interface would need to provide a way to specify an ID. One easy way to enable this search is to allow a user to initiate a search on an entity by clicking on it in a document, e.g., clicking on *AT&T* in Figure 1 would start a search for other documents that mention this specific company. Alternatively, such mentions of names in text could be hyperlinked to renderings of the corresponding database information. This landing page might have links out to other documents mentioning the entity. Another use would be to gather entity-based summary information such as how many cases has Microsoft been a defendant in this past year or what the law firms were, defending AT&T in the U.S. first circuit federal court system.

In the remainder of this article we present a number of methods for recognizing named entities and describe a system based on these methods. We then discuss the resolution task and again describe an actual system.

2   Named Entity Recognition

We use three methods for NE recognition: lookup, pattern rules, and statistical models. These methods can also be combined in hybrid systems.

The lookup method consists of creating a list of names of the entities of interest, such as drug names, and then simply tagging all mentions of elements in the list as entities of the given type. For example, if Vioxx is in the list of names and it appears in a document, then mark it as a drug name. Drug names are often unusual and thus often unambiguous: if they appear in text, then, with a high degree of certainty, the words refer to the drug. In contrast, a list of judge names would contain many names that are common in the greater population and thus not unambiguously judge names. Often such common names can be weeded out to ensure that the list reaches an acceptable level of ambiguity. For example, using U.S. government census data, the commonness of names can be estimated. The advantages of the lookup approach are that it is simple to implement and maintain, it can make use of preexisting lists, it will find names regardless of context, and it does not require any training data. The disadvantages are that it may generate many false positives,[1] if the list contains many ambiguous words.

---

[1] A false positive is a situation where the NE recognizer believes that a name exists but it does not. For example, a lookup tagger might propose that *Bush* refers to George W. Bush in *My Life in the Bush of Ghosts was recorded in 1981 with Brian Eno*. A false negative is a name that is passed over by the tagger, e.g., perhaps *Brian Eno* in the sentence above. True positives are those names that the tagger finds and true negatives are those non-names that are correctly ignored by the tagger. Two measures of accuracy can be defined using counts of these classifications. Precision is the number of true positives divided by the sum of the counts of true and false positives. Recall is true positives divided by the sum of true positives and false negatives. Precision measures how often the guessed entities are correct and recall measures how often the system finds the entities that are actually there.

It may also make a number of false negatives if the list is not comprehensive enough. The basic problem is that lookup taggers ignore contextual cues for names and their types. An example cue would be that person names usually follow *Mr.*

Contextual rules encode such cues in deductive rules, e.g., if a capitalized word sequence follows *Mr.*, tag it as a person name. By looking at development data, a knowledge engineer can develop a set of such rules that recognizes the majority of the instances in the data and does not produce many false positives. The advantages of this approach is that often quite high accuracy can be obtained. The disadvantages are that the development of such rules requires manually annotated development data and often a large amount of effort from experienced rule writers. The development data is required so that the rule writers have examples to build rules from, examples to test new rules, and examples to do regression testing against. In addition to having enough manually annotated development data, it is also important for this data to be representative: ideally it should be a random sample of the data that the system will see at runtime. Maintenance of such rule sets can be a challenge, since often the rules have intricate inter-dependencies that are easy to forget and make modification risky. Finally, many descriptions of cues involve words such as "usually" or "sometimes". Providing some sort of weighting on the rules can encode such partial aspects. However, how to set these weights optimally is often a difficult and time consuming task and further complicates maintenance.

Statistical models offer an alternative to contextual rules for encoding contextual cues. One way of thinking about such statistical models is as a set cues that receive weights and whose weights are combined based on probability and statistical concepts. A knowledge engineer must develop features that correspond to cues, pick the appropriate statistical model, and train the model using training data. As with contextual rules, statistical models can achieve high accuracy. The disadvantages are that the development of such models requires manually annotated training data and often a large amount of effort from an experienced machine learning expert. The training data is the same sort of data as the development data mentioned above: it needs to be representative of the data that the model will process at runtime. In our experience, the amount of manually annotated data required for writing good contextual rules is very similar to the amount of data needed to train a good statistical model. However, maintenance can be easier than with contextual rules in that only more hand annotated data is required to change a model's behavior.

Each of these three approaches has its place in the toolkit of an engineer trying to build named entity recognizers for legal text, and we will see examples of each in the system we describe below. In addition, the methods can be combined in a number of ways. See [7] for a discussion of how best to combine such systems. We need to point out that we did not invent the three methods above; in fact, each has a long history in the named entity recognition literature. There has been a wealth of work done on named entity recognition in the natural language processing community but we will not try to review it here nor provide an exhaustive bibliography. Instead we will direct the reader to the following review articles: review of work on newswire text [9], review of work in bio-health domain [8]. For what concerns legal texts, see [1,3] and [4].

2.1   A Named Entity Recognition System for Legal Text

Legal documents can be classified as captioned/non-captioned. Captioned documents have extensive header material preceding the body of the document, which usually includes attorney information, court & jurisdiction, parties (plaintiffs and defendants), and document title. Captions may also list the primary judge of the case and an identification number for the case. There are also litigation documents without the caption information but we will focus on captioned documents here. Figure 2 shows a sample captioned document. All named entity recognition systems described in this section except the judge tagger work with captioned documents.

Preprocessing. Before a captioned document is processed by the named entity tagger, we apply the following preprocessing : tokenization, zoning, and line-blocking.

Tokenization is the process of breaking up text into its constituent tokens. Our tokenization also involves identifying the line breaks in the text. Court, judge, and jurisdiction taggers work at the token level and the others use lines as input.

The zoning task involves identifying the caption portion of the legal document. Our zoner uses a statistical model (a conditional random field) [10] which was trained on a manually-tagged training set of around 400 documents. The features used can be classified into the following categories:

- N-gram features: frequent n-grams occurred around the zone in the training data,
- Positional features: position of the line in the document like the first quarter, last quarter, first-tenth etc.,
- Punctuation features: punctuation like dashed lines are good indicators of the beginning of a new zone.

Line blocking is the process of identifying structural blocks of text in a document. It groups contiguous lines of the document into meaningful blocks. We developed a rule-based system based on formatting and case information to identify the blocks.

Jurisdiction Tagger. Jurisdiction refers to a particular geographic area containing a defined legal authority. Jurisdiction can be Federal, State or even smaller geographical areas like cities and counties. Jurisdiction is usually tied to the court. In order to find the jurisdiction, the jurisdiction tagger performs a longest substring match on the court context. Court context is the 5 line window surrounding the first line in the document containing the word *court*.

Court Tagger. Court tagger is a lookup tagger. It extracts different components of the court like the jurisdiction, division and circuit from the court context and looks for a court in the authority database with those constituents.

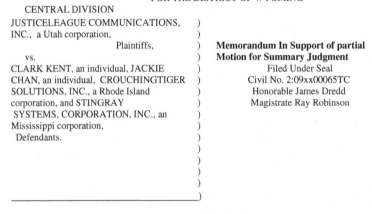

Adam B. Eve (2312)
Max M. Payne (2721)
Swinburne Wilde & Larkin
201 Main Street, Suite 23
Salt Lake City, Utah 84321-1219
Telephone: (801) 555-5555
Facsimile: (801) 555-5556
Ming Chang*
Joseph Stalin*
Lincoln Jefferson & Roosevelt, PC
One Cross Street
Oconomowoc, Wisconsin 34233-3423
Telephone: (334) 555-2009
Facsimile: (334) 555-2002
Attorneys for Defendants Clark Kent,
Jackie Chan and CrouchingTiger Solutions, Inc.
*Admitted Pro Hac Vice

IN THE UNITED STATES DISTRICT COURT
FOR THE DISTRICT OF WYOMING

CENTRAL DIVISION

| JUSTICELEAGUE COMMUNICATIONS, INC., a Utah corporation,
Plaintiffs,
vs.
CLARK KENT, an individual, JACKIE CHAN, an individual, CROUCHINGTIGER SOLUTIONS, INC., a Rhode Island corporation, and STINGRAY SYSTEMS, CORPORATION, INC., an Mississippi corporation,
Defendants. |))))))))))))))) | **Memorandum In Support of partial Motion for Summary Judgment**
Filed Under Seal
Civil No. 2:09xx00065TC
Honorable James Dredd
Magistrate Ray Robinson |

Introduction

It is important in any trade secrets case to determine whether the Plaintiff owns any trade secrets and, if so, to define the scope of such secrets. Defendants Jackie Chan, Clark Kent and CrouchingTiger Solutions, Inc. file this Partial Motion for Summary Judgment because Plaintiff JusticeLeague Communications, Inc. has admitted on Rule 32(b)(7) deposition that is does not have any admissible evidence that the GBear Code that it alleges was copied

Fig. 2. A captioned legal document

Title Tagger. We used a statistical approach to extract titles from the document. A title can span more than one line in the document. So as a predecessor to the title tagger, we apply line blocking to group adjacent lines into blocks. The title classifier classifies each block as title or not. Features for the title classifier include case information, formatting, position, and length. We used a manually annotated data set of 400 documents to train the classifier.

Doctype Tagger. A legal document can be assigned to a category based on its contents (its doctype). Some example doctypes are Brief, and Memorandum.

The doctype is usually a part of the title of the document. Once the title of the document is found, the doctype tagger performs a longest substring match on the title to find the doctype from the doctype list.

Judge Tagger. The judge tagger looks for honorifics such as *Honorable, Hon., Judge* in the document and extracts names that follow them as judges. We also developed a similar attorney tagger. These taggers are prototypical context rule taggers.

Table 1 lists the precision and recall for all the taggers. The test set includes 600 documents randomly selected from a large collection of legal documents.

Table 1. Precision and Recall of Taggers

Tagger	Precision	Recall
Jurisdiction	97	87
Court	90	80
Title	84	80
Doctype	85	80
Judge	98	72

2.2   Summary of Named Entity Recognition

In this section, we have introduced the named entity recognition task and described three approaches to performing it: lookup, context rules, and statistical models. We then presented a system that recognizes a number of named entities types in captioned legal documents. In the next section, we move on to named entity resolution.

3   Named Entity Resolution

The job of entity resolution is to map the names with class labels to particular entities in the real world by assigning them to entity records within a class authority file.

A class authority file is a file in which each record identifies a particular person or entity in the real world belonging to the class. Examples of authority files pertinent to the legal domain include files for attorneys, judges, expert witnesses, companies, law firms, and courts.

As mentioned above resolution enables a number of applications. Thomson West's Profiler is an example of an application that allows hyperlinking from names in text to text pages describing the named entity [1]. In this system, attorneys, judges, and expert witnesses in caselaw documents, dockets, pleadings, and briefs are tagged and linked to the curriculum vitae of particular legal professionals. A screen shot showing a caselaw document with attorneys and judges highlighted is shown in Figure 3.

Briefs and Other Related Documents

United States Court of Appeals,
First Circuit.
Diane DENMARK, Plaintiff, Appellant,
v.
LIBERTY LIFE ASSURANCE COMPANY OF BOSTON, and The Genrad, Inc. Long Term Disability Plan, through Teradyne, Inc
as Successor Fiduciary, Defendants, Appellees.

No. 05-2877.
July 2, 2008.

Jonathan M. Feigenbaum, Phillips & Angley, Boston, MA, for Plaintiff, Appellant.

Ashley B. Abel, Jackson Lewis, Greenville, SC, Andrew C. Pickett, Jackson Lewis, Maia M. Rafik, S. Stephen Rosenfeld,
Rosenfeld & Rafik, Boston, MA, Eugene R. Anderson, Anderson Kill & Olick, New York, NY, Amy Bach, Law Offices of Amy
Bach, Mill Valley, CA, Daniel J. Healy, Rhonda D. Orin, Anderson Kill & Olick, Jay E. Sushelsky, American Association of
Retired Persons, Litigation Dept., Carolyn Doppelt Gray, Teresa L. Jakubowski, Barnes & Thornburg LLP, Washington, DC, f
Defendants, Appellees.

Fig. 3. Caselaw document displayed in Westlaw with attorney names highlighted. Users
may jump directly to attorney's curriculum vitae by clicking on name.

The Profiler application also shows how NE resolution enables indexing of
all references to particular entities within text collections [2]. In Profiler, a user
can find all the cases, briefs, and settlements in which a particular attorney has
been mentioned. An example of the index display for an attorney is shown in
Figure 4.

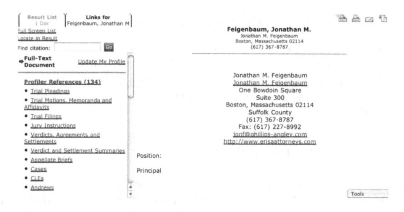

Fig. 4. Profiler page for attorney Jonathan Feigenbaum. Page shows curriculum vitae of
Mr. Feigenbaum and also lists all cases, briefs, motions, pleadings, and other documents
mentioning Mr. Feigenbaum.

An example of a system that is built upon extracted relationships between
resolved entities is Thomson West's Litigation Monitor [3]. For the Litigation
Monitor, we extracted and resolved judges, attorneys, and companies in caselaw
documents and dockets throughout the federal and state legal systems. We took
the additional step of linking the company references to their representing attorneys. By combining these resolutions, Litigation Monitor can show all the
companies a particular attorney has represented in federal and state court as

well as all the attorneys a particular company has used for litigation. By combining this information with firm, court, judge, legal topic, time information, and role information (i.e., is a company the defendant or the plaintiff?), Litigation Monitor serves as a robust utility that can display litigation trends across multiple informational axes. Figure 5 shows all the law firms that have represented the Microsoft Corporation in federal and state court.

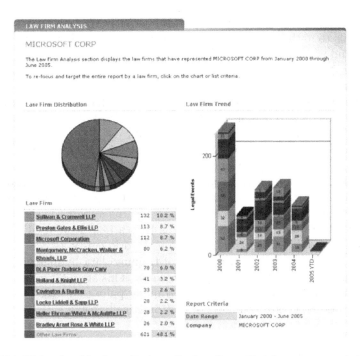

Fig. 5. This Litigation Monitor displays the law firms that have represented the Microsoft Corporation in federal and state courts between the year 2000 and the beginning of 2005

3.1   Record Linkage Approach to Named Entity Resolution

The mapping to an authority file approach to NE resolution involves a two step process: information extraction and record linkage. The extraction part consists of moving a tagged name from a given semantic class and information co-occurring with the name into a structured record. The record linkage part involves matching the structured record to a record within an authority file.

The subtasks we need to undertake to build a NE resolution system include the following.

The first task is to analyze the authority files' inherent ambiguity. The purpose of this task is to determine which sets of record attributes uniquely identify a record in the authority file.

The second task consists of designing a text structured record that includes enough fields to, in most cases, enable an unambiguous match to the authority. A text structured record is a structured record that contains information about a named entity tagged in text. For example, a text structured record for an attorney in a caselaw record would include the first, middle and last name of the attorney taken from the tagged attorney name in text and it might also include names of cities, states, and law firms that co-occur in the same paragraph with the attorney name.

The third task involves manually matching a relatively small set of text structured records to authority file records. This set of record pairs is called the test match set and is used as a gold standard against which to measure precision and recall of the NE resolution system. We also use the test set to evaluate record blocking strategies.

The fourth task is to design a blocking strategy for retrieval of candidate authority file records to match against a text structured record. We use the test match set to guide the blocking strategy selection.

The fifth task consists of developing a set of match feature functions to compare candidate authority file record fields to text structured record fields. The output of these feature functions will be used to create a match feature vector. The matching classifier accepts as input the match feature vector associated with an authority-text structured record pair and returns a match belief score. The match belief score indicates how likely it is that a text structured record and authority file record pair refer to the same real world entity.

The sixth task involves creating a set of training data by matching a set of text structured records to authority file records. This step is similar to step 2 except that we need a larger number of training pairs than test pairs, and in some cases we can create training data automatically using surrogate training features.

The seventh task is to train a match belief scorer using the training data. The scorer can be based on a variety of machine learning techniques. In this paper, we will discuss the use of a support vector machine to train a scorer for matching attorney names in case law to an attorney authority file.

And the final task is to deploy structured record creation, blocking, feature function processing, and match belief scoring in a pipe line to resolve text entities tagged in text to authority file records. We assign entity ids from authority record to text names to create text hyperlinks and indexes.

In the following sections, we discuss each of these steps and illustrate them with the example of matching attorney names in caselaw to an authority file of attorney records. In this application, we tagged the attorney names, cities, state, and law firm names in 43,936 U.S. federal cases. From the representation paragraphs in these cases, we extracted text structured records for each attorney name and matched the attorney names to an authority file of U.S. attorneys. A representation paragraph is a paragraph in which the attorneys representing the parties to the case are listed. A representation paragraph is shown in Figure 6. The text structured record for *Mark D. Stubbs* is shown in Table 2.

Mark D. Stubbs, Barnard N. Madsen, Matthew R. Howell, Fillmore Spencer LLC, Provo, UT, Ashby D. Boyle, II, Fillmore Spencer, Sandy, UT, for Plaintiffs.

Fig. 6. Example of Legal Representation Paragraph

Table 2. Example Text Structured Record

Record Field Label	Field Value
first name	Mark
middle name	D
last name	Stubbs
name suffix	null
city-state name	Provo,UT:Sandy,UT
firm name	Fillmore Spencer LLC

3.2   Analyzing Ambiguity of Authority File

The ambiguity of an authority file means the degree to which the entities in the authority file have overlapping identifying information. For example, if a large percentage of the attorneys in the attorney authority file have the same first and last name, then the ambiguity of this file would be high with respect to these two attributes.

A useful way to assess the ambiguity of a file is to measure the mean number of records that share the same set of attribute settings for any given record in the file. A large mean number of records for a particular set of attribute values means that this set of attributes on its own is not likely to produce good resolution accuracy. A mean cluster size very close to 1.0 for a set of attributes, on the other hand, is likely to produce good resolution accuracy (provided the authority file is fairly well populated).

One can discover viable attribute sets to use for name resolution matching by measuring the ambiguity of different combinations of attributes.

Table 3 shows an analysis of the ambiguity of different attribute combinations for attorneys in our attorney file. The attorney authority file lists over one million U.S. attorneys. The identifying attributes for an attorney include first, middle, and last name as well as city, state, and firm name.

We can see from Table 3 that on average a given first and last name combination is associated with 4.02 authority file records. So a record linkage system that relied only on these fields to make a match would not be very accurate. On the other hand, combining first and last name with firm, location, or middle name information would yield accurate results. Note that we would expect record linkage accuracy to text structured records to be lower than accuracy predicted by our ambiguity analysis because information in the text structured record may be incorrect due to extraction errors or may be out of sync with information in the authority, as when an attorney changes firms or even last names due to marriage. However, our ambiguity analysis does show what kind of accuracy we

Table 3. Mean Number of Records Sharing Field Set Attributes in Attorney Authority File

field sets	mean number records
first+last	4.0193
first+last+state	1.2133
first+last+city+state	1.0718
first+last+firm	1.0454
first+last+firm+state	1.0175
first+last+middle	1.000

could expect with perfect extraction and synchronization of data between text and authority file.

3.3   Designing Text Structured Record

The text structured record is a record whose fields are populated with information extracted from text that pertains to a particular tagged named entity. The fields of the text structured record should align semantically with the fields in the authority file. The set of fields in the text structured record should encompass enough information that an unambiguous match can be made to an authority file record when the text fields are populated. The text structure record typically will include the name of the tagged entity as well as other co-occurring information that corresponds to the attributes in the authority file record. Our analysis of the ambiguity of the authority file can indicate to us which sets of attributes we should incorporate into our text structure record.

3.4   Developing Test Data

To test an NE resolution system, we need a set of pairs of authority-text structured records that refer to the same individual (positive instances) and a set of pairs of authority-text structure records that do not refer to the same individual (negative instances). We measure the precision of our NE resolution system by dividing the number of correct positive matches made by the total number of matches made. We measure recall by dividing the number of correct positive matches made by the number of positive instances manually identified.

Since the authority file should contain at most one record that truly matches a named reference in text, we can automatically create many negative test examples from each positive authority-structured record pair by assuming that all authority-structured record pairs derived from a retrieved block of candidate records are negative (i.e. do not co-refer) except the positive (i.e. co-referring) pair which has been identified.

One method of creating test data is to randomly select structured records extracted from the text and manually match them to records in the authority file. Typically, we want to have at least 300 positive test record pairs. For our example attorney NE resolution system, we created 500 manual matches between

Table 4. Blocking Efficiency for Three Different Block Keys

block key	recall	mean block size
last name	1.0	403.2
last name+first init	0.97	7.6
last name+first name	0.95	4.4

attorney names mentioned in our database of U.S. federal cases and attorneys listed in our attorney authority file.

3.5   Selecting Blocking Functions

The purpose of blocking is to increase the efficiency of an NE resolution system by limiting the number of authority records to which we must compare the text structured record in order to find a match. A good blocking function will return a small mean number of authority records for text structured records under consideration and will also return with a high probability the authority record that best matches each text structured record.

We can use our test data to assess different blocking functions. In Table 4, we show the blocking efficiency of three different blocking keys. To compute these numbers, we constructed a text structured record from text associated with each of our 500 test instances. We then counted the number of records returned from the authority file when we used each of the three lookup keys: last name, last name+first initial of first name, and last name+first name. We also counted how often the returned block of authority records contained the matching authority record that had been manually identified. We computed the mean block size by dividing the number of records returned for each key by the number of test instances (i.e., 500). We computed the recall (the probability that block contains matching record) by dividing the number of block sets that contained the manually identified authority record by the number of test instances.

3.6   Selecting Matching Feature Functions

Match features are functions that compare semantically compatible fields from the authority file record and the structured record in such a way that the more closely the argument field values match the higher the feature function value.

The operation of particular feature functions is governed by the semantics of the fields being compared. Two types of similarity functions merit special mention because they work well for many field classes. They are the TFIDF cosine similarity function and the family of string edit distance functions.

The TFIDF cosine similarity function computes the cosine similarity of a string field in the text structure record and a string field value in the authority record. The inverse document frequency values for words in the record field are computed from the field in the authority file. Each record in the authority file is considered a document and the number of different records in which a word appears in the string field is used as the word's occurrence count [5].

The family of string similarity functions includes among others the Hamming distance, Levenshtein distance, Smith-Waterman distance, and Jaro-Winkler distance. Each of these functions measures the distance between strings as a function of the number of character changes that need to be made to transform one string into another [5].

For our example NE resolution system, we use the following five similarity functions to convert an attorney authority file and a text structured record pair into a feature vector. In our example, we block using last name, so we need no feature function for last name. In our Profiler system, we do incorporate a last name feature function and use a blocking function other than simple last name. For purposes of illustration, however, using last name as the blocking function works adequately.

First name similarity function. This compares the first name of the attorney field in the authority file with the first name in the text structured record. The function returns a value of 1.0 if the names match exactly, a value of 0.8 if the names match as nick names, a value of 0.8 if one name is only specified by an initial and matches the first letter of the other name, a value of 0.5 if one or both of the first names is unspecified, and a value of 0.0 if the names mismatch.

Middle name similarity function. This compares the middle name or initial of the attorney field in the authority file with the middle name or initial in the text structured record. This function is essentially the same as the first name similarity function except it is applied to middle names.

Name suffix similarity function. This feature function compares the name suffix value (e.g., "Jr." or "III") in the authority file record with the name suffix value in the text structured record. This function returns a 1.0 if both suffixes are specified and match, a 0.8 if both suffixes are unspecified, a 0.5 if a suffix is specified in one name and not in the other, and a 0.0 if the suffixes are specified and mismatch.

City-state similarity function. This feature compares city-state information in the authority file record with city-state information in the text structured record. The function returns a 1.0 if both city and state match, a 0.75 if just the state matches, a 0.5 if city-state information is not specified in one or both records, and a 0.0 if the states mismatch. Note that we consider all city-state pairs that co-occur in the same paragraph with the attorney name and use the highest scoring city-state pair to set our feature function value.

Firm name similarity function. This feature measures the TFIDF cosine similarity between the law firm name specified in the authority file record and law firm names in the text structured record. Note that we consider all firm names that co-occur in the same paragraph with the attorney name reference and use the highest scoring firm name to set our feature function value.

Note that these feature functions are somewhat simpler than the feature set we use for Profiler. However, these functions work well here to illustrate NE resolution concepts.

3.7   Developing Training Data

To train an NE resolution record pair classifier, you need a relatively large set of authority-text structured record pairs that co-refer (positive data) and a set of authority-text structure record pairs that do not co-refer (negative data). The positive and negative training data are used by a machine learning program to create a model that combines feature function values to yield a match belief score for a given feature function vector associated with a authority-text structured record pair.

The amount of training data one needs for a NE resolution system varies according to the nature of names, text, and authority file associated with the system. Our experience has been that a few thousand positive and negative record pairs works fairly well.

Training data can be created manually or in some cases automatically. To create training data manually, we follow the same procedure we used to create test data. To create training data automatically, we pick one or two feature fields that under certain conditions can be used to very probably identify a positive match in a candidate record block. We call the feature fields that can automatically identify positive matches surrogate features. Note that negative training data can be generated for NE resolution automatically from positive training instances in the same way they were generated for the test data.

The surrogate feature method of automatically creating positive training data works well if there are non-surrogate features that are independent of the surrogate features and that are robust enough to identify matching records within the candidate authority record block. The attraction of using surrogate features is that we do not have to incur the cost of manually creating training data. The drawback of using surrogate features is that, once we have used them to identify positive matches, we cannot reuse them as match features in our scoring program.

For our example NE resolution system, we created 5,000 manually matched positive training examples.

We also created 70,000 positive instances automatically in our caselaw corpus by using rare first and last name combinations as a surrogate feature. We selected rare names by identifying first and last name pairs in census data that occur fewer than 50 times in the general U.S. population and only once in the attorney authority file. Any attorney file and text file record pairs that have a matching rare first-last name combination are considered positive matches.

3.8   Using Support Vector Machine to Learn Matching Function

A good method of linking structured records to authority records is to use a Support Vector Machine (SVM) classifier that measures the likelihood that a particular structured record and authority record refer to the same entity. An advantage of the SVM is that it can learn non-linear functions of feature values relatively easily.

Table 5. Precision, Recall and F-measure of the NE resolution system using manual and surrogate training data to train SVM matching function

Training data	Precision	Recall	F-measure
Manual	0.96	0.95	0.95
Surrogate	0.96	0.89	0.92

To create an NE resolution system using an SVM, we provide the SVM with positive and negative training instances to obtain a model of support vectors with which to score feature function vectors. The higher the score the model returns for a feature function the higher the likelihood that the authority file record and text structured record pair refer to the same person.

For our NE resolution example system, we created one matching model from the manual training data and one model from the automatic training data.

3.9   Assembling and Testing the NE Resolution Pipeline

We createt an NE resolution system by assembling our text structured record generator, blocking function, feature vector generator, and feature vector scorer into a pipeline that will find a matching authority record for a given text name reference. The pipeline functions as follows for a given text reference:

1. Create a text structured record for a text name reference.

2. Create blocking key from the structured record. This value in our tests was the last name of the attorney name tagged in text.

3. Retrieve all records from the authority file that match the blocking key. These are the authority candidate records.

4. Pair each authority candidate record with the current text structure record and compute a feature function vector.

5. Score each feature function vector with the trained SVM.

6. Choose the authority file record associated with the highest scoring feature vector as the best match to the text structure record.

7. If the highest scoring feature vector exceeds a minimum threshold, assign the entity id associated with the matching authority file record to the text reference associated with the text structure record.

We measured how well the SVM approach works using manual and automatically acquired training data for the attorney names in case law. Our results are shown in Table 5.

The f-measure using our surrogate training approach was only three percentage points below our manual training approach. The difference was the six percentage point drop in recall. We suspect this drop was due to the fact that some matching attorneys instances involved first names expressed as nick names or initials and our surrogate training examples excluded these types of matches.

Overall however the surrogate approach using rare names was fairly effective. We have discussed using rare names for surrogate training more fully in [6].

4   Conclusion

One aspect of semantic processing of legal texts is figuring out what people, courts, companies, law firms, etc. are mentioned in a text. This processing can be broken into two parts: (i) recognizing names of certain types and (ii) resolving these names to specific individuals in the world. We have outlined both of these tasks and presented systems for performing them in the legal domain.

References

1. Dozier, C., Haschart, R.: Automatic Extraction and Linking of Person Names in Legal Text. In: Proceedings of RIAO 2000, Recherche d'Information Assistee par Ordinateur, Paris, France, April 12-14, pp. 1305–1321 (2000)
2. Dozier, C., Zielund, T.: Cross Document Co-Reference Resolution Applications for People in the Legal Domain. In: Proceedings of the ACL 2004 Workshop on Reference Resolution and its Applications, Barcelona, Spain, July 25-26, pp. 9–16 (2004)
3. Chaudhary, M., Dozier, C., Atkinson, G., Berosik, G., Guo, X., Samler, S.: Mining Legal Text to Create a Litigation History Database. In: Proceedings of IASTED International Conference on Law and Technology, Cambridge, MA, USA (2006)
4. Quaresma, P., Gonçalves, T.: Using Linguistic Information and Machine Learning Techniques to Identify Entities from Juridical Documents. In: Francesconi, E., Montemagni, S., Peters, W., Tiscornia, D. (eds.) Semantic Processing of Legal Texts. LNCS (LNAI), vol. 6036, pp. 44–59. Springer, Heidelberg (2010)
5. Cohen, W., Ravikumar, P., Fienberg, S.: A Comparison of String Distance Metrics for Name-matching Tasks. In: Proc. II Web Workshop IJCAI, pp. 73–78 (2003)
6. Dozier, C., Veeramachaneni, S.: Names, Fame, and Co-Reference Resolution, Thomson Reuters Research and Development. Technical Report (2009)
7. Liao, W., Light, M., Veeramachaneni, S.: Integrating High Precision Rules with Statistical Sequence Classifiers for Accuracy and Speed. In: Proceedings of the NAACL 2009 Workshop Software engineering, testing, and quality assurance for Natural Language Processing (2009)
8. Yeh, A., Morgan, A., Colosimo, M., Hirschman, L.: BioCreative task 1A: Gene mention finding evaluation. BMC Bioinformatics 6(Suppl. 1) (2005)
9. Grishman, R., Sundheim, B.: Message Understanding Conference - 6: A Brief History. In: Proceedings of the 16th International Conference on Computational Linguistics (COLING), I, Kopenhagen (1996)
10. Lafferty, J., McCallum, A., Pereira, F.: Conditional random fields: Probabilistic models for segmenting and labeling sequence data. In: Proc. of ICML (2001)

Using Linguistic Information and Machine Learning Techniques to Identify Entities from Juridical Documents

Paulo Quaresma and Teresa Gonçalves

Departamento de Informática, Universidade de Évora
7000-671 Évora, Portugal
{pq,tcg}@di.uevora.pt

Abstract. Information extraction from legal documents is an important and open problem. A mixed approach, using linguistic information and machine learning techniques, is described in this paper. In this approach, top-level legal concepts are identified and used for document classification using Support Vector Machines. Named entities, such as, locations, organizations, dates, and document references, are identified using semantic information from the output of a natural language parser. This information, legal concepts and named entities, may be used to populate a simple ontology, allowing the enrichment of documents and the creation of high-level legal information retrieval systems.

The proposed methodology was applied to a corpus of legal documents - from the EUR-Lex site – and it was evaluated. The obtained results were quite good and indicate this may be a promising approach to the legal information extraction problem.

Keywords: Named Entity Recognition, Natural Language Processing, Machine Learning.

1 Introduction

Information extraction from text documents is an important and quite open problem, which is increasing its relevance with the exponential growth of the "web". Every day new documents are made available online and there is a need to automatically identify and extract their relevant information.

Although this is a general domain problem, it has a special relevance in the legal domain. For instance, it is crucial to be able to automatically extract information from documents describing legal cases and to be able to answer queries and to find similar cases.

Many researchers have been working in this domain in the last years, and a good overview is done in Stranieri and Zeleznikow's book "Knowledge Discovery from Legal Databases" [1]. Proposed approaches vary from machine learning techniques, applied to the text mining task, to the use of natural language processing tools.

E. Francesconi et al. (Eds.): Semantic Processing of Legal Texts, LNAI 6036, pp. 44–59, 2010.
© Springer-Verlag Berlin Heidelberg 2010

We propose a mixed approach, using linguistic information and machine learning techniques. In this approach, top-level legal concepts are identified and used for document classification using a well known machine learning technique – Support Vector Machines. On the other hand, named entities, such as, locations, organizations, dates, and document references, are identified using semantic information from the output of a natural language parser. The extracted information – legal concepts and named entities – may be used to populate a simple ontology, allowing the enrichment of documents and the creation of high-level legal information retrieval systems. These legal information systems will have the capacity to retrieve legal documents based on the concepts they convey or the entities referred in the texts.

The proposed methodology was applied to a corpus of legal documents from the EUR-Lex site[1] within the "International Agreements" sections and belonging to the "External Relations" subject. The obtained results were quite good and they indicate this may be a promising approach to the legal information extraction problem.

The paper is organised as follows: section 2 describes the main concepts and tools used in our approach – SVM for text classification and a syntactic/semantic parser for named entities recognition – and the document collection used to evaluate the proposal; section 3 describes the experimental setup for the identification of legal concepts task and evaluates the obtained results; section 4 describes the named entity recognition task and its results; section 5 briefly describes some related work; and, finally, section 6 presents some conclusions and points out possible future work.

2   Concepts and Tools

This section introduces the concepts and tools employed in this work: the machine learning text classification approach used to automatically identify legal concepts and the appliance of linguistic information for named entity recognition. It concludes by presenting the exploited juridic dataset.

2.1   Text Classification

The learning problem can be described as finding a general rule that explains data, given a sample of limited size. In supervised learning, we have a sample of input-output pairs (the *training sample*) and the task is to find a deterministic function that maps any input to an output such that the disagreement with future input-output observations is minimised. If the output space has no structure except whether two elements are equal or not, we have a *classification* task. Each element of the output space is called a *class*. The supervised classification task of natural language texts is known as *text classification*.

In text classification, documents must be pre-processed to obtain a more structured representation to be fed to the learning algorithm. The most common

---

[1] http://eur-lex.europa.eu/en/index.htm

approach is to use a bag-of-words representation, where each document is represented by the words it contains, with their order and punctuation being ignored. Normally, words are weighted by some measure of word's frequency in the document and, possibly, the corpus. Figure 1 shows the bag-of-words representation for the sentence "The provisions of the Agreement shall be applied to goods exported from South Africa to one of the new Member States.".

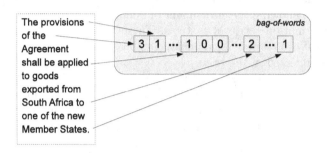

Fig. 1. Bag-of-words representation

In most cases, a subset of words (stop-words) is not considered, because their role is related to the structural organisation of the sentences, and does not have discriminating power over different classes. Some work reduces semantically related terms to the same root applying a lemmatiser.

Research interest in this field has been growing in the last years. Several machine learning algorithms were applied, such as decision trees [2], linear discriminant analisys and logistic regression [3], the naïve Bayes algorithm [4] and Support Vector Machines (SVM)[5].

[6] says that using SVMs to learn text classifiers is the first approach that is computationally efficient and performs well and robustly in practice. There is also a justified learning theory that describes its mechanics with respect to text classification.

Support Vector Machines. Support Vector Machines, a learning algorithm introduced by Vapnik and coworkers [7], was motivated by theoretical results from the statistical learning theory. It joins a kernel technique with the structural risk minimisation framework.

Kernel techniques comprise two parts: a module that performs a mapping from the original data space into a suitable feature space and a learning algorithm designed to discover linear patterns in the (new) feature space. These stages are illustrated in Figure 2.

The *kernel function*, that implicitly performs the mapping, depends on the specific data type and domain knowledge of the particular data source.

The *learning algorithm* is general purpose and robust. It's also efficient, since the amount of computational resources required is polynomial with the size and number of data items, even when the dimension of the embedding space (the feature space) grows exponentially [8].

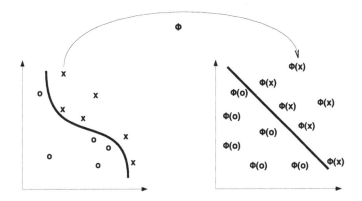

Fig. 2. Kernel function: data's nonlinear pattern transformed into linear feature space

Four key aspects of the approach can be highlighted as follows:

- Data items are embedded into a vector space called the feature space.
- Linear relations are discovered among the images of the data items in the feature space.
- The algorithm is implemented in a way that the coordinates of the embedded points are not needed; only their pairwise inner products.
- The pairwise inner products can be computed efficiently directly from the original data using the kernel function.

The *structural risk minimisation* (SRM) framework creates a model with a minimised VC (Vapnik-Chervonenkis) dimension. This developed theory [9] shows that when the VC dimension of a model is low, the expected probability of error is low as well, which means good performance on unseen data (good generalisation).

In geometric terms, it can be seen as a search to find, between all decision surfaces (the $\mathcal{T}$-dimension surfaces that separate positive from negative examples) the one with maximum margin, that is, the one having a separating property that is invariant to the most wide translation of the surface. This property can be enlighten by Figure 3 that shows a 2-dimensional problem.

SVM can also be derived in the framework of the regularisation theory instead of the SRM one. The idea of regularisation, introduced by [10] for solving inverse problems, is a technique to restrict the (commonly) large original space of solutions into compact subsets.

Classification Software. SVM^{light} [11] is a Vapnik's Support Vector Machine [12] implementation in C^2. It is a fast optimization algorithm [6] that has the following features:

---

[2] Available at `http://svmlight.joachims.org/`

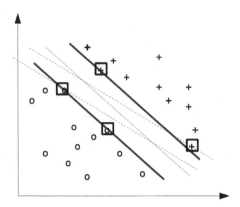

Fig. 3. Maximum margin: the induction of vector support classifiers

- solves classification, regression and ranking problems [13]
- handles many thousands of support vectors
- handles several hundred-thousands of training examples
- supports standard kernel functions and lets the user define your own

SVM^{light} can also train SVMs with cost models [14] and provides methods for assessing the generalization performance efficiently, the XiAlpha-estimates for error rate and precision/recall [15,6].

This tool has been used on a large range of problems, including text classification [16,5], image recognition tasks, bioinformatics and medical applications. Many of these tasks have the property of sparse instance vectors and using a sparse vector representation, it leads to a very compact and efficient representation.

2.2   Named Entity Extraction

A named entity extractor locates in text the names of people, places, organizations, products, dates, dimensions and currency. This information is needed to complete the final step in formation extraction of populating the attributes of a template. It is also useful to locate sentences that contain particular entities to answer questions.

To address this task machine learning techniques such as decision trees [17], Hidden Markov Models [18] and rule based methods [19] have been applied. In this work, instead of using a statistical approach, we will use a linguistic one.

Linguistic Information. The written language has a specific structure and comprehends several information levels. The most simple ones are the morphological and syntactic ones.

Morphological information includes a word's stem and its morphological features, like grammatical class and inflectional information. While some natural language processing tasks use a word's stem, others use its lemma.

Most syntactic language representations are based on the context-free grammar (CFG) formalism introduced by [20] and, independently, by [21]: given a sentence, it generates the corresponding syntactic structure. It is usually represented by a tree structure, known as sentence's *parse tree*, that contains its constituent structure (such as noun and verb phrases) and the grammatical class of the words.

Syntactic Parser Tool. Documents' syntactic structure was obtained using the PALAVRAS [22] parser for the English language. This tool was developed in the context of the VISL project by the Institute of Language and Communication of the University of Southern Denmark[3].

Given a sentence, the output is a parse tree enriched with some semantic tags. This parser is robust enough to always give an output even for incomplete or incorrect sentences, which might be the case for the type of documents used in text classification, and has a comparatively low percentage of errors (less than 1% for word class and 3-4% for surface syntax) [23].

For example, the output generated for the sentence "The provisions of the Agreement shall be applied to goods exported from South Africa to one of the new Member States." is

```
STA:fcl
=SUBJ:np
==>N:art("the" S/P) The
==H:n("provision" <act> <sem-c> <ss> <nhead> <left> P NOM) provisions
==N<:pp
===H:prp("of" <np-close>) of
===P<:np
====>N:art("the" S/P) the
====H:n("agreement" <sem-c> <act-s> <ss> <ac-cat> <nhead> S NOM) Agreement
=P:vp
==VAUX:v-fin("shall" <aux> PR) shall
==VAUX:v-inf("be" <aux>) be
==MV:v-pcp2("apply" <mv> PAS) applied
=PIV:pp
==H:prp("to" <right>) to
==P<:np
===H:n("goods" <cc-h> <nhead> P NOM) goods
===N<:icl
====P:v-pcp2("export" <mv> <np-close> PAS) exported
====ADVL:par
=====CJT:pp
======H:prp("from" <cjt-head> <advl-close> <right>) from
======P<:n("South_Africa" <complex> <nhead> <Proper> <Lcountry> S NOM) South_Africa
====P<<:pp
=====H:prp("to") to
=====P<:adjp
======H:num("one" <card> S) one
======N<:pp
=======H:prp("of" <np-close>) of
=======P<:np
=========>N:art("the" S/P) the
========>N:adj("new" POS) new
========H:n("member_States" <complex> <nhead> <Proper> <heur> S NOM) Member_States
.
```

---

[3] Available at http://www.visl.sdu.dk/

2.3   Dataset Description

We performed the experiments over an set of European Union law documents. These documents were obtained from the **EUR-Lex** site[4] within the "International Agreements" section, belonging to the "External Relations" subject matter.

From all available agreements we chose the ones that had their full text (not just the bibliographic notice) and obtained a set of 2714 agreements dating from 1953 to 2008. Since the agreements are available in several languages we collected them for two anglo-saxon languages (English and German) and for two romanic ones (Italian and Portuguese), obtaining four different corpora: eurlex-EN, eurlex-DE, eurlex-IT and eurlex-PT.

Table 1 presents, for each corpus, the total number and average per document of tokens (running words) and types (unique words).

Table 1. Total number and average per document of tokens and types for each corpus

corpus	tokens total	per doc	types total	per doc
eurlex-EN	10699234	3942	73091	570
eurlex-DE	10145702	3728	133191	688
eurlex-IT	10665455	3929	96029	636
eurlex-PT	9731861	3585	86086	567

Each eurlex document is classified according to several ontologies: one obtained using the "EUROVOC descriptor", other using the "Directory code" and another using the "Subject matter". In all available classifications each document can be assigned to several categories. This setting is known as "multi-label".

The identification of legal concepts was accomplished using the first level of the "Directory code" classification, considering only the categories with at least 50 documents. Table 2 shows each category (id and name) along with the number of documents assigned to it.

Table 2. Number of documents assigned to each category

id	name	# of docs
2	Customs Union and free movement of goods	209
3	Agriculture	390
4	Fisheries	361
7	Transport policy	81
11	External relations	2628
12	Energy	58
13	Industrial policy and internal market	55
15	Environment, consumers and health protection	138
16	Science, information, education and culture	99

---

[4] Available at http://eur-lex.europa.eu/en/index.htm

3   Legal Concepts Identification

This section introduces the experimental setup and presents and evaluates the results obtained for the legal concepts identification task.

3.1   Experimental Setup

The experiments were done using a bag-of-words representation of documents, the SVM algorithm was run using SVM^{light} with a linear kernel and other default parameters and the model was evaluated using a 10-fold stratified cross-validation procedure.

Document Representation. To represent each document we used the bag-of-words approach, mapping all numbers to the same token and using the tf-idf weighting function normalised to unit length. This well known measure weights word w_i in document d as

$$\text{tf-idf}(w_i, d) = tf(w_i, d) \ln \frac{N}{df(w_i)}$$

where $tf(w_i, d)$ is the w_i word frequency in document d, $df(w_i)$ is the number of documents where word w_i appears and N is the number of documents in the collection.

Stratified Cross-validation. The cross-validation (CV) is a model evaluation method where the original dataset is divided into k subsets (in this work, $k = 10$), each one with (approximately) the same distribution of examples between categories as the original dataset (stratified CV). Then, one of the k subsets is used as the test set and the other k-1 subsets are put together to form a training set; a model is built from the training set and then applied to the test set. This procedure is repeated k times (one for each subset). Every data point gets to be in a test set exactly once, and gets to be in a training set $k - 1$ times. The variance of the resulting estimate is reduced as k is increased.

Performance Measures. To measure learner's performance we analysed precision, recall and the F_1 measures [24] of the positive class. These measures are obtained from the contingency table of the classification (prediction *vs.* manual classification). For each performance measure we calculated the micro- and macro-averaging values of the top ten categories.

Precision is the number of correctly classified documents (true positives) divided by the number of documents classified into the class (true positives plus false positives).

Recall is given by the number of correctly classified documents (true positive) divided by the number of documents belonging to the class (true positives plus false negatives).

F_1 is the weighted harmonic mean of precision and recall and belongs to a class of functions used in information retrieval, the F_β-measure. F_β can be written as follows

$$F_\beta(h) = \frac{(1 + \beta^2)prec(h)rec(h)}{\beta^2 prec(h) + rec(h)}$$

Macro-averaging corresponds to the standard way of computing an average: the performance is computed separately for each category and the average is the arithmetic mean over the ten categories.

Micro-averaging does not average the resulting performance measure, but instead averages the contingency tables of the various categories. For each cell of the table, the arithmetic mean is computed and the performance is computed from this averaged contingency table.

All significance tests were done regarding a 95% confidence level.

3.2   Results

While Figure 4 shows the micro- and macro-average precision, recall and F_1 graphically, Table 3 shows those measures for each category. For each measure, micro- and macro-average boldface values have no significant difference between them and the best value obtained.

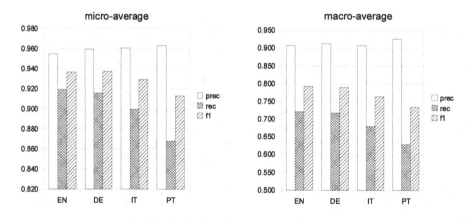

Fig. 4. Micro- and macro-average values

3.3   Evaluation

As can be seen in Figure 4, the precision values are good and the same for all studied languages (there's no significant difference between them): the micro-precision is above 0.95 while the macro one is above 0.90.

Having smaller values, the recall measure does not present the same behaviour: the best micro and macro-recall is for the English corpus, with .919 and .721 respectively, but while for the micro measure there is no significant difference

Table 3. Precision, recall and F_1 values for each category

id	eurlex-EN			eurlex-DE			eurlex-IT			eurlex-PT		
	prec	rec	F_1	prec	rec	F_1	prec	rec	F_1	prec	rec	F_1
2	.907	.651	.758	.952	.665	.783	.903	.579	.706	.929	.565	.702
3	.914	.818	.863	.926	.805	.861	.939	.705	.805	.942	.503	.656
4	.955	.934	.944	.965	.906	.934	.979	.914	.946	.971	.823	.891
7	.821	.568	.672	.846	.543	.662	.792	.519	.627	.813	.481	.605
11	.973	.998	.985	.973	.997	.985	.973	.998	.985	.973	.997	.985
12	.949	.638	.763	.872	.707	.781	.886	.672	.765	.921	.603	.729
13	.913	.382	.538	.895	.309	.459	.889	.291	.438	.944	.309	.466
15	.901	.725	.803	.918	.732	.815	.909	.725	.806	.902	.732	.808
16	.837	.778	.806	.868	.798	.832	.899	.717	.798	.941	.646	.766
micro	**.955**	**.919**	**.937**	**.960**	**.916**	**.937**	**.961**	**.900**	**.929**	**.964**	.868	.913
macro	**.908**	**.721**	**.792**	**.913**	**.718**	**.790**	**.908**	.680	.764	**.926**	.629	.734

for the German and Italian languages, for the macro one only the anglo-saxon languages present the best values.

Considering the individual category results, it is possible to conclude that the precision is always above recall for all languages and categories and as expected (since documents where retrieved having the "External relations" subject matter), the "External relations" category (id 11) have the best precision and recall with values almost equal to one in all languages. The "Fisheries" (id 4) also have very good values all above .9 (except the recall for the Portuguese corpus).

On the other way, there are some categories with small recall:

- while "Industrial policy and internal market" (id 13) has the worst ones, with values between .309 for the Portuguese corpus and .382 for the English one,
- "Transport policy" (id 7) has values between .481 for the Portuguese corpus and .568 for the English one and
- "Customs Union and free movement of goods" (id 2) and "Energy" (id 12) have values between .565 ("Customs" category for the Portuguese language) and .707 ("Energy" category for the German corpus).

Comparing results between languages, the English and German corpus present the best and very similar results, with the Portuguese one presenting the worst ones.

4   Named Entity Recognition

This section presents the experiments done for Named Entity Recognition. It begins by describing the experimental setup, then the results are presented and an evaluation is made.

4.1   Experimental Setup

The experiments were done using the `eurlex-EN` corpus (the collection for the English language). The following categories of Named Entities were studied:

- location names
- organization names
- dates
- references to documents and document articles

We did not try to extract personal names since after analysing the corpus we found almost no references to them.

For the extraction of location names and organization names we used the following subset of the semantic tags given by the parser PALAVRAS (see section 2.2):

- `<Lwater>`, `<Ltown>`, `<Lregion>` and `<Lcountry>` for location names
- `<HHorg>` and `<comp2>` for organization names

For the identification of dates we used a simple NLP tool, which received as input the sentences parse tree and performed a tree match procedure able to identify dates. References to other article and documents were also identified from the analysis of the parse trees.

After obtaining the candidate Named Entities, and since the corpus was not tagged, a manual evaluation was made for each category. For location names we made the analysis using the categorization given by PALAVRAS: "water" names (oceans, seas, rivers, etc. . .), towns, regions and countries.

4.2   Results

Table 4 shows for each kind of extracted named entities, the number of documents and for tokens (running words) and types (unique words) the total number and the minimum, maximum and average per document.

It is important to point out that we didn't obtain the number of unique references because we only identified and extracted the references inside the documents and we didn't try to consolidate the results. In order to be able to calculate this value we will need further text processing and it will be the focus of future work.

Table 4. Number of documents and for tokens (running words) and types (unique words) the total number and the minimum, maximum and average per document (for each kind of named entity)

category	docs	tokens				types			
		total	min	max	avg	total	min	max	avg
water	180	964	1	206	5.36	56	1	20	1.81
town	1820	11981	1	2001	6.58	307	1	54	2.32
region	1075	19438	1	456	18.08	220	1	46	2.77
country	2142	63979	1	621	29.87	521	1	97	4.72
organization	2281	56571	1	568	24.80	70	1	19	2.98
date	2714	19994	1	–	7.36	3521	1	–	1.29
reference	2714	76091	0	–	28.03	–	–	–	–

Table 5 presents the error percentage for each kind of named entity studied.

Table 5. Error percentage for each kind of named entity

category	error
water	12.5%
town	13.7%
region	18.2%
country	28.2%
organization	67.1%
date	0.1%
reference	65%

4.3   Evaluation

From table 4 we can state that these documents have a high number of references to other documents and articles (76091 references found and a 28% average per document). They also have high values of references to organizations and countries (56571 and 63979, respectively). These values are compatible with the type of analysed documents: legislation from the European Union. They also help to support our claim that this kind of information extraction is very important and it would allow the inference of important relations, such as, the chain of legislation references.

19994 date references were also identified, related to 3521 distinct events. This information can also be used as a basis for an analysis of relevant events in this legislation domain.

The performed evaluation focused on the precision of the information extraction modules and the results were shown in table 5. There are 3 classes of results:

- dates – The precision was quite good (error rate of 0.1%). This precision value was obtained because the legal documents have a quite standard way of presenting dates and a simple NLP tool was able to identify and extract the dates;
- location – Precision between 80 and 90%. These results depend heavily on the quality of the semantic tag classifier of the parser. We observed typical classes of errors and a simple upgrade of the parser's geographical information should significantly improve these results;
- organization and references – Precision around 35%. This quite low value has distinct explanations:
  - organization – the problem is caused by the semantic tag classifier of the parser. From a preliminary analysis it seems that all entities unknown to the system are classified as "organization". Only a change in the parser will allow an improvement of this result. Another approach might be to develop a special SVM classifier for this kind of entities.
  - reference – The high error rate value is explained by the complex syntactic structure used in the documents to make references to articles of other legislation. A deeper analysis of the syntactic sentence structure is needed to improve the quality of this sub-task.

5   Related Work

As referred in section 1 much work has been done in this domain in the last years. A good overview is done in the Stranieri and Zeleznikow's book "Knowledge Discovery from Legal Databases" [1]. In this book several approaches to the legal information extraction problem are described, varying from machine learning techniques to natural language processing methodologies. A more general but relevant reference in the information extraction domain is the "Information Extraction" paper of J. Cowie and W. Lehnert [25].

In the legal domain some of the related work is:

- [26] used decision trees to extract rules to estimate the number of days until the final case disposition;
- [27] developed rule based and neural networks legal systems;
- [28] used neural networks to model legal classifiers;
- [29,30] used SVM to classify juridical Portuguese documents;
- [31] proposed a framework for the automatic categorisation of case laws;
- [32,33] described the use of self-organising maps (SOM) to obtain clusters of legal documents in an information retrieval environment and explored the problem of text classification in the context of the European law;
- [34] described classification and clustering approaches to case-based criminal summaries;
- [35,36,37] described also related work using linear classifiers for documents;
- [38] integrated information extraction, information retrieval and machine learning techniques in order to design a case-based retrieval system able to find prior relevant cases. They used SVMs to rank prior case candidates.

6   Conclusions and Future Work

A proposal to identify and extract concepts and named entities in legal documents was presented and evaluated. The proposed methodology uses a SVM classifier to associate concepts to legal documents and a natural language parser to identify named entities, namely, locations, organizations, dates, and references to other articles and documents.

The concept classification task obtained an precision higher than 0.95 for the four languages selected in this experience (English, German, Italian, and Portuguese). Worst results were obtained for the romanic languages, which is compatible with previous research and is probably due to the use of more complex syntactic structures and richer morphology.

The named entities task obtained very good results for the identification of dates, an average result for locations (10-20% average error rate) and bad results for the identification of organizations and references to other articles and legislation. Extraction of locations can improve with the use of geographical databases and with the availability of this information to the parser – this will be the focus of future work. The identification of references to other articles and legislation needs a deeper analysis of the parse trees: from our error analysis we were able to conclude that further work needs to be done in order to fully understand these syntactic structures.

Finally, we will improve our legal information retrieval system [39,40] to take into account the extracted information and to allow users to retrieve documents based on semantic information and not on surface-level words.

References

1. Stranieri, A., Zeleznikow, J.: Knowledge Discovery from Legal Databases. In: Law and Philisohy Library. Springer, Heidelberg (2005)
2. Tong, R., Appelbaum, L.: Machine learning for knowledge-based document routing. In: Harman (ed.) Proceedings of the 2nd Text Retrieval Conference (1994)
3. Schütze, H., Hull, D., Pedersen, J.: A comparison of classifiers and document representations for the routing problem. In: SIGIR 1995, 18th ACM International Conference on Research and Developement in Information Retrieval, Seattle, US, pp. 229–237 (1995)
4. Mladenić, D., Grobelnik, M.: Feature selection for unbalanced class distribution and naïve Bayes. In: ICML 1999, 16th International Conference on Machine Learning, pp. 258–267 (1999)
5. Joachims, T.: Transductive inference for text classification using support vector machines. In: ICML 1999, 16th International Conference on Machine Learning (1999)
6. Joachims, T.: Learning to Classify Text Using Support Vector Machines. Kluwer Academic Publishers, Dordrecht (2002)
7. Cortes, C., Vapnik, V.: Support-vector networks. Machine Learning 20, 273–297 (1995)
8. Shawe-Taylor, J., Cristianini, N.: Kernel Methods for Pattern Analysis. Cambridge University Press, Cambridge (2004)
9. Vapnik, V.: Statistical learning theory. Wiley, New York (1998)
10. Tikhonov, A., Arsenin, V.: Solution of Ill-Posed Problems. John Wiley and Sons, Washington (1977)
11. Joachims, T.: Making large-scale SVM learning practical. In: Schölkopf, B., Burges, C., Smola, A. (eds.) Advances in Kernel Methods - Support Vector Learning. MIT Press, Cambridge (1999)
12. Vapnik, V.: The nature of statistical learning theory. Springer, New York (1995)
13. Joachims, T.: Optimizing search engines using clickthrough data. In: Schölkopf, B., Burges, C., Smola, A. (eds.) KDD 2002, 8th ACM Conference on Knowledge Discovery and Data Mining. ACM, New York (2002)
14. Morik, K., Brockhausen, P., Joachims, T.: Combining statistical learning with a knowledge-based approach – A case study in intensive care monitoring. In: ICML 1999, 16th International Conference on Machine Learning (1999)
15. Joachims, T.: Estimating the generalization performance of a SVM efficiently. In: Schölkopf, B., Burges, C., Smola, A. (eds.) ICML 2000, 17th International Conference on Machine Learning. MIT Press, Cambridge (2000)
16. Joachims, T.: Text categorization with support vector machines: Learning with many relevant features. In: ECML 1998, 10th European Conference on Machine Learning, Chemnitz, DE, pp. 137–142 (1998)
17. Baluja, S., Mittal, V., Sukthankar, R.: Applying Machine Learning for High Performance Named Entity Extraction. Computational Intelligence 16 (2000)

18. Miller, S.: Nymble: a high-performance learning name-finder. In: ALNP 1997, 5th Conference on Applied Natural Language Processing, pp. 194–201 (1997)
19. Aberdeen, J., Burger, J., Day, D., Hirschman, L., Robinson, P., Vilain, M.: MITRE: description of the Alembic system used for MUC-6. In: MUC6 1995, 6th Conference on Message Understanding, Morristown, NJ, USA, pp. 141–155. Association for Computational Linguistics (1995)
20. Chomsky, N.: Three models for the description of language. IRI Transactions on Information Theory 2, 113–124 (1956)
21. Backus, J.: The syntax and semantics of the proposed international algebraic of the Zurich ACM-GAMM Conference. In: Proceedings of the International Conference on Information Processing – IFIP Congress, UNESCO, Paris, pp. 125–132 (1959)
22. Bick, E.: The Parsing System PALAVRAS – Automatic Grammatical Analysis of Portuguese in a Constraint Grammar Framework. Aarhus University Press (2000)
23. Bick, E.: A constraint grammar based question answering system for portuguese. In: Pires, F.M., Abreu, S.P. (eds.) EPIA 2003. LNCS (LNAI), vol. 2902, pp. 414–418. Springer, Heidelberg (2003)
24. Salton, G., McGill, M.: Introduction to Modern Information Retrieval. McGraw-Hill, New York (1983)
25. Cowie, J., Lehnert, W.: Information extraction. Commun. ACM 39, 80–91 (1996)
26. Wilkins, D., Pillaipakkamnatt, K.: The effectiveness of machine laerning techniques for predicting time to case disposition. In: ICAIL 1997, 6th International Conference on Artificial Intelligence and Law, pp. 39–46. ACM, New York (1997)
27. Zeleznikow, J., Stranieri, A.: The split-up system: Integrating neural networks and rule based reasoning in the legal domain. In: ICAIL 1995, 5th International Conference on Artificial Intelligence and Law, pp. 194–195. ACM, New York (1995)
28. Borges, F., Borges, R., Bourcier, D.: Artificial neural networks and legal categorization. In: Proccedings of the 16th International Conference on Legal Knowledge Based Systems, pp. 11–20. IOS Press, Amsterdam (2003)
29. Gonçalves, T., Quaresma, P.: A preliminary approach to the multilabel classification problem of Portuguese juridical documents. In: Pires, F.M., Abreu, S.P. (eds.) EPIA 2003. LNCS (LNAI), vol. 2902, pp. 435–444. Springer, Heidelberg (2003)
30. Gonçalves, T., Quaresma, P.: Is linguistic information relevant for the classification of legal texts? In: Sartor, G. (ed.) Proceedings of the 10th International Conference on Artificial Intelligence and Law, Bologna, Italy, pp. 168–176. ACM, New York (2005)
31. Thompson, P.: Automatic categorization of case law. In: ICAIL 2001, 8th International Conference on Artificial Intelligence and Law, pp. 70–77 (2001)
32. Schweighofer, E., Merkl, D.: A learning technique for legal document analysis. In: ICAIL 1999, 7th International Conference on Artificial Intelligence and Law, pp. 156–163. ACM, New York (1999)
33. Schweighofer, E., Rauber, A., Dittenbach, M.: Automatic text representation, classification and labeling in european law. In: ICAIL 2001, 8th International Conference on Artificial Intelligence and Law, pp. 78–87 (2001)
34. Liu, C.L., Chang, C.T., Ho, J.H.: Classification and clustering for case-based criminal summary judgement. In: ICAIL 2003, 9th International Conference on Artificial Intelligence and Law, pp. 252–261 (2003)
35. Brüninghaus, S., Ashley, K.D.: Improving the representation of legal case texts with information extraction methods. In: ICAIL 2001: Proceedings of the 8th international conference on Artificial intelligence and law, pp. 42–51. ACM, New York (2001)

36. Brüninghaus, S., Ashley, K.: Finding factors: learning to classify case opinions under abstract fact categories. In: ICAIL 1997, 6th International Conference on Artificial Intelligence and Law, pp. 123–131. ACM, New York (1997)
37. Brüninghaus, S., Ashley, K.D.: Predicting outcomes of case-based legal arguments. In: ICAIL 2003, 9th International Conference on Artificial Intelligence and Law, pp. 233–242 (2003)
38. Al-Kofahi, A.K., Vachher, A., Jackson, P.: A machine learning approach to prior case retrieval. In: ICAIL 2001, 8th International Conference on Artificial Intelligence and Law, pp. 88–93 (2001)
39. Quaresma, P., Rodrigues, I.: A question-answering system for legal information retrieval. In: Moens, M.F., Spyns, P. (eds.) Proceedings of JURIX 2005: The 18th Annual Conference on Legal Knowledge and Information Systems, Frontiers in Artificial Intelligence and Applications, Brussels, Belgique, pp. 91–100. IOS Press, Amsterdam (2005)
40. Quaresma, P., Rodrigues, I.: A question-answering system for portuguese juridical documents. In: Sartor, G. (ed.) Proceedings of the 10th International Conference on Artificial Intelligence and Law, Bologna, Italy, pp. 256–257. ACM, New York (2005)

Approaches to Text Mining Arguments from Legal Cases

Adam Wyner[1], Raquel Mochales-Palau[2], Marie-Francine Moens[2],
and David Milward[3]

[1] Department of Computer Science
University College London
Gower Street, London WC1E 6BT, United Kingdom
adam@wyner.info
www.wyner.info/research
[2] Computer Science Department
Katholieke Universiteit Leuven
Afdeling Informatica, Celestijnenlaan, 3001 Heverlee, Belgium
raquel.mochalespalau, sien.moens@cs.kuleuven.be
www.cs.kuleuven.be/cs/
[3] Linguamatics Ltd
St. John's Innovation Centre, Cowley Rd
Cambridge, CB4 0WS, United Kingdom
david.milward@linguamatics.com
www.linguamatics.com

Abstract. This paper describes recent approaches using text-mining to automatically profile and extract arguments from legal cases. We outline some of the background context and motivations. We then turn to consider issues related to the construction and composition of corpora of legal cases. We show how a Context-Free Grammar can be used to extract arguments, and how ontologies and Natural Language Processing can identify complex information such as case factors and participant roles. Together the results bring us closer to automatic identification of legal arguments.

Keywords: Text Mining, Legal Argument, Case–based Reasoning.

1 Introduction

In countries with legal systems using common law, such as the United States and the United Kingdom, case law plays a critical role in legal reasoning and decision-making. Case law is that corpus of decisions on cases which judges have made; we refer to this corpus as the *case base* and the previous decisions as the *precedents*. Given a current case, a lawyer consults the case base and identifies precedents that support their side in the legal dispute and undermine the other. The lawyer presents the precedents as, in effect, analogical arguments: the precedents are related to the current case in certain respects, and as the precedent was decided, so too should the current case. However, the cases are expressed in natural language, consider highly complex matters that are under dispute, relate to laws which justify the decision, and have complex inter-relationships

E. Francesconi et al. (Eds.): Semantic Processing of Legal Texts, LNAI 6036, pp. 60–79, 2010.
© Springer-Verlag Berlin Heidelberg 2010

such as when one case decision overturns a prior case decision. Legal professionals must undergo very extensive training in navigating the case base, interpreting the results, and applying the results successfully to their current case. Adding to this complexity, the case base is comprised of a large number of cases and grows every year[1]. Thus, the legal professional faces the difficult task of retrieving and interpreting information from the case base.

Historically, legal professionals have a variety of tools they have been able to use to manage and search the case base in order to identify the relevant cases and material (e.g. compilations of decided cases as well as *Shepard's Citations*, which indexes cases with respect to applicable precedents). More recently, with electronic documentation and automated techniques, legal professionals can search the case base quickly and with respect to a range of parameters. Large companies such as *Lexis-Nexis* and *Westlaw* provide legal information along with access to legal case bases to legal professionals. As we discuss below, the information allows legal professionals to search through the case base relative to a set of terms and quickly returns a set of candidate cases for the legal professional to consider. However, while there are tools to refine the search, by and large, the results returned are fairly coarse-grained; it is up to the legal professional to read the case abstract or the body of the case itself to determine if it suits the case at hand.

Automated text mining tools that can perform information extraction on the case base have a range of advantages. Using such tools, detailed properties and relationships within and among cases can be identified. Searches can be carried out and information can be made available to legal researchers on new cases automatically as they are added to the case base. The goal of information extraction is to automatically extract structured information from unstructured machine-readable texts. The information is structured in that it identifies semantic properties or relationships in the texts. For example, suppose we want to identify the set of lawyers who have never lost a case in a certain domain (e.g. real estate) along with the set of cases the lawyers used to argue their side. Information extraction is more specific than information retrieval, which identifies documents, rather than lists of lawyers or relationships between lawyers and successful cases. For example, search engines such as *Google* identify sources using keywords and return sets of links in which the keywords are found, but not the relationships among the keywords.

The paper has several objectives. First, it is of significant research interest to automatically identify legal arguments, properties, and relationships as found in legal cases. Second, the research here can be viewed as a contribution to the development of tools which support legal professionals in their activities such as identifying relevant cases in the case base. Finally, the research can help us better understand the meaning of the law and the functioning of the legal system by, in effect, a bottom-up investigation starting from case law, which is the foundation of the common law system.

---

[1] See searchable databases available on website of the World Legal Information Institute (WorldLii) with links to databases of legal decisions of countries such as the USA:
http://www.worldlii.org/
Also see the databases of US law at the Legal Information Institute of the Cornell Law School:
http://www.law.cornell.edu/

The paper is structured as follows. In section 2, we outline the relevant background and scope of our work such as the range of approaches to argumentation and previous related work. In section 3, we discuss issues about the development of a corpus of legal arguments as it appears in recent research. A variety of approaches have been taken, and we indicate some of the advantages and disadvantages of each. Section 4 presents current results of three approaches to information extraction in legal cases. In 4.1, the focus is on parsing arguments into an *argument interchange format* as well as to try to automatically identify argument sentences from non-argument sentences. In 4.2, the emphasis is identifying argument characteristics from a set of legal documents and providing a context-free grammar. In 4.3, a range of semantically relevant elements are extracted from a case base which include those indexed by commerical services, but also novel complex information such as case factors. Together the approaches provide highly related aspects of the automatic identification of legal arguments wherein participants argue about issues such as case factors. In 5, we discuss the relation of this work with that of others as well as provide indications of the direction of future work. We use the terms "argumentation" and "argument", where argumentation is about the abstract theory and argument relates to particular instances that may be chained together; for example, we have theories of argumentation, while *if Bill is unhappy, then he should leave the party. If Bill leaves the party, then he should take his dog with him* is a chain of arguments (see [1] for related discussion).

2 Background

In this section, we outline the range of approaches to argumentation in order to set the context of our work. In addition, we refer to some key prior work in the area of legal text mining.

2.1 Argumentation Theory and Analysis

Research in argumentation is interdisciplinary, relating discussions found in Philosophy, Linguistics, and Computer Science. *Empirically* oriented approaches attend to specific, linguistically realised argument structures, properties, or elements of legal texts ([2], [3], [4], [5], [6], and [7]). Given analyses of argument patterns, arguments can be graphically represented as trees, where premises branch off of conclusions [8]. XML markup languages have been developed for argumentation such that an argument, once marked up, can be searched for or used for reasoning [9]. While the results of some of this work can be used for information extraction, it is not produced automatically and is not suitable for working with large corpora.

In computational models of argumentation, the abstract structures of and reasoning with arguments are proposed. In Argumentation Frameworks, arguments are abstract and atomic objects in attack relations ([10], [11], [12], [13], and [14]). Argumentation Frameworks account for a range of issues in non-monotonic reasoning [15]; they can be extended to accommodate more fine-grained elements of relations between arguments such as different modes of attack ([14], [1], and [16]). Such approaches focus on high level generalisations about sets of arguments and the complexity of their relationships. However, the theories do not address a range of aspects of natural argumentation.

There have been attempts to connect abstract arguments with concrete arguments ([17], [18], [19], and [20]). Such systems start with a knowledge base comprised of facts and rules, where the rules typically include both strict (*SI*) and defeasible (*DI*) inference rules in a Defeasible Logic (*DL*) ([21] and [22]). In these approaches, simplified examples are manually translated into the formal language and the objective is to draw inferences given the knowledge base. An additional realisation of argumentation theory are the argument schemes for case based reasoning of [23], which relates arguments and case based reasoning by allowing schemes to be used to argue for or against a decision given comparative case factors found in the case base.

2.2   Text-Mining in the Case Base

Current commerical systems (e.g. Lexis-Nexis or Westlaw) or web-based public services (e.g. WorldLii) have limited text mining capabilities. One way to identify a set of relevant cases is by selecting from among a small finite set of indicis, which are manually assigned to the cases. More 'advanced' facilities support regular expression searches along with boolean operators and proximity operators. Such facilities do not reflect any semantic information.

A range of other approaches search for semantic content. Recent work in computational semantics focuses on recognising inference, calculating entailment, and identifying inconsistency ([24], [25], [26], [27], [28], [29] and [30]). Such work does not address arguments with complex structure, relationships between arguments, or key elements of legal information found in the case base. [31] presents a representation of knowledge of cases that would suit an information retrieval system for a case base, outlining different sorts of knowledge that ought to be identified and by which cases are classified – functional, structural, semantic, and factual. Some systems provide information retrieval of case factors in order to support case based reasoning ([32], [33], and [34]). However, they do not report how text mining is used to identify the factors, nor whether the approach could support queries beyond those specifically designed for the particular case based reasoner. [35] extracts cases in an appellate chain, which are those cases that are relevant to the current case in terms of the comments on the quality of the case, e.g., whether it has been appealed, affirmed, overturned, overruled, explained, or distinguished. [35] focus on automated extraction of citation relations, not on argumentation or case factors. [36] parses individual sentences from legal texts. However, the results do not bear on parsing arguments nor on information extraction. [37] and [38] develop text mining approaches to identify the *rhetorical* structures in free texts; where the argument indicators are explicit, the rates of identification are reasonably high, but fall where the indicators must be inferred. See [39] for a survey of other systems and approaches to information retrieval from legal texts as of the 1990s.

2.3   Summary

We have briefly surveyed a number of different approaches to argumentation and information extraction from case bases. In our approach we have two angles of attack on these issues. First, we are concerned with information extraction of argument structures from the case base, allowing us to identify decisions and their justifications. Second, we want to extract not only key profile elements (along the lines of the functional elements of [31]), but also the linguistically represented case factors that can be used in case based reasoning (along the lines of [32] and [34]).

3 Argument Corpus

In carrying out text mining, the first task is to form a corpus from which the information will be extracted. In this section, we discuss several previous attempts to form argument corpora, citing the data sources and purposes of the corpora. We also point out several additional potentially useful sources of arguments.

3.1 Araucaria

[40] outlines the creation of an argument database *AraucariaDB*. In the current version, *AraucariaDB* contains approximately 700 arguments. The arguments are drawn from a range of international sources. There are several problems concerning the sampling. We are not given a criteria which was used to guide the selection of the arguments. No consideration is given to the impact of the different source contexts: the arguments found in a newspaper may be different from those found in judicial summaries; in addition, arguments in Japan may be different from those in the United States. By the same token, there is no information concerning how arguments were identified from the source material, thus there could be biases of arguments of a certain sort or on a particular topic, namely whatever was perceived by the selector to have been an argument. While subjective criteria need not be problematic, the absence of overt criteria hinders evaluation of the resulting selection. Indeed, it is not clear what supports the claim that the contents of the database are arguments. In sum, given the small sample, lack of context information, and lack of criteria, it is uncertain what we can infer about the patterns which may appear.

3.2 Mochales and Moens

The corpus in [41] contained 30 relevant documents. The documents were divided into development and test documents. The 10 development documents were used to construct a grammar and to establish a *gold standard*. They were legal decisions drawn from an online database of cases from the European Court of Human Rights (ECHR), which has a common law legal system. All the texts were independently marked by three parties, who came to agreement about the mark up; however, there are interesting observations concerning differences among the markers concerning the identification of implicit premises or the extent of portions of arguments. It is reasonable to assume that the cases contain legal arguments, unlike documents sourced elsewhere. The documents contain a variety of formal sections such as statements of facts, complaints, and the reasons for and against the decision. As [41] note, not all these sections are expected to contain arguments; for example, a statements of the facts, precedents, or procedural moves might not be taken as samples of argument. The 20 test documents were used to test the adequacy of the argument grammar.

3.3   Wyner and Milward

[42] provide two corpora of 50 and 90 legal cases selected arbitrarily from a search of the British and Irish Legal Information Institute's (BAILII) online database of legal decisions in the United Kingdom and Ireland. Within the BAILII database, a keyword search was made for cases pertaining to medical malpractice so as to have a coherent set of cases. As with the ECHR cases, it is reasonable to assume that the cases contain legal arguments, and broadly contain reports of facts, complaints, and the reasons for or against the decision, as well as applicable law. One objective of the study was to develop text mining tools to automatically search for elements that are found in commercial case law search engines, such as indices for citation index, judges, jurisdiction, and so on. Another objective was to develop searches for features of the case beyond those found in such search engines, such as case features or the identification of violation of some norm. Together these aspects of cases are crucial for argumentative case based reasoning [23].

3.4   Others

In addition to the sources of argument indicated above, there are a range of other available resources which ought to be considered in future research. For example, *Debateopedia* is an online encyclopedia of debates, which are arguments pro and con a range of particular issues. It is compiled under the auspicies of the *International Debate Education Association*, so it has a claim to be representing arguments as they are expressed.[2] The structure of the arguments, providing arguments pro and con an issue, is closer to what is understood to be an argument than inference patterns. The database contains several thousand debates from which a selection can be made concerning a particular topic of interest. For the legal domain, the *World Legal Information Institute* is a searchable index of databases of case law including links to UK and US case law databases.[3] Finally, the US *National Institute of Standards and Technology* has been sponsoring a task for *recognising textual entailment*, in which natural language processing techniques are applied to a database of entailment patterns.[4] Discussion on the selection and analysis of the database is discussed in [28].

4   Analysis

In analysing a case base, a variety of approaches have been applied. In 4.1, the arguments identified in ArcauriaDB were manually marked up in an XML which indicates the structure of the argument. In 4.2, a context free grammar for arguments is proposed then used to identify and parse an argument from a case. Such an approach promises to allow the extraction of arguments of a specific structure from the case base. In 4.3, cases in a case base are profiled with respect to a range of functional features with text mining tools; in addition, relational factors such as failure to fulfil and obligation are identified.

---

[2] http://wiki.idebate.org/index.php/Welcome_to_Debatepedia!

[3] http://www.worldlii.org/

[4] http://www.nist.gov/tac/2009/RTE/index.html

4.1 Araucaria

In [40], the arguments in the corpus were analysed and represented in an XML-based format, the *Argument Markup Language* (AML).[5] AML represents arguments in terms of XML markups that indicate a range of properties and relationships among the propositions that constitute the argument. Argument schemes, which are stereotypical patterns of reasoning, are used to guide and catalog the arguments [5]. Suppose an argument from the argument from sign scheme *A bear passed this way. Here are some bear tracks in the snow.* We infer from the evidence of the bear tracks that a bear presumably passed by. AML indicates the scheme, the distinct propositions, the relation between them (one as the premise and the other at conclusion) as well as a range of auxiliary information such as date of analysis and author of analysis. More complex arguments can be represented in AML. The XML format is machine-readable, so can be searched for argument components; in addition, it supports translation into a graphical representation, which can be easily understood.

For the analysis of the AraucariaDB, two analysts applied AML using argument schemes. No systematic methodology is outlined, and there were no controls for intercoder reliability. As with the development of the corpus, the methodology of analysis makes it difficult to draw reliable conclusions about the data.

Another approach to the analysis of the AraucariaDB is found in [37]. The objective of the study is to be able to automatically identify the argument from non-argument sentences drawn from the corpus. The AraucariaDB is augmented with a similar number of sentences which are claimed to be non-argumentative. The features used to distinguish examples are: sequences of words in a sentence (from one (unigrams) to three (trigrams)), words identified by part-of-speech (adverbs, verbs, and modal auxiliaries), collocations (high word pair co-occurrences), text statistics (sentence length, word length, punctuation marks), punctuation, parse features (given a parse of the sentence into phrases, the depth of the tree and number of subclauses), and key words that have been identified as signalling argument (e.g. *but*, *consequently*, *because of*, and others).

The best results were obtained with a combination of collocations, verbs, sentence length, word length, and number of punctuation marks. The latter three broadly can be taken to indicate that sentence complexity (length of sentence and words along with complex structure signaled by punctuation) signal argument.

The results are limited in several respects. First, no criteria is given for the selection of the non-argument sample sentences; it is not clear to what extent sentence complexity itself signals argument or whether this is an artifact of how argument and non-argument sentences were selected for the corpus. Second, the role of key words and collocations is unclear; just what key words or collocations signal argument is not specified. Moreover, as [37] note, some features are ambiguous; the modal auxiliaries in particular have a so-called *root* and *epistemic* interpretation, where only the former might be used to infer an argument [44]. Finally, again as [37] note, arguments contain *enthymemes*, which are implicit reasoning steps [45]; thus, if an argument must be inferred from information not found in the text, it cannot be identified by text mining techniques.

---

[5] This seems largely superseded by the *Argument Interchange Format* in [43].

4.2   Mochales and Moens

[41] focus on argument identification within a limited corpus and relative to a gold standard as well as the formulation of a context-free argument grammar. In this section, the grammar is proposed, applied, and evaluated.

Grammars. Formalisation of natural languages using grammars has been a topic of interest among linguistic researches for years ([46], [47], and [48]). Modern computational notations given by grammars can be used to develop parsing applications. Context Free Grammars (CFG) have been used extensively for defining the syntax of a variety of natural and artificial (e.g. programming) languages [49]. In this section, we formally present a CFG for legal arguments, exploiting the inherently structured nature of argument in case law documents, specifically in the judgements and decisions of the European Court of Human Rights (ECHR). We develop a parsing application to test the grammar accuracy.

CFGs are a particular class of grammar, where grammars are composed of a finite set of terminal and non-terminal symbols, a special start symbol, and a finite set of productions, which are rules of substitution whereby the left-hand symbols are substituted for by the right-hand symbols. CFGs allow substitutions independent of the rest of the structure; the left-hand side of rule can only consist of a single symbol and the right-hand side is a non-empty string over the total vocabulary, i.e. terminal and non-terminal symbols. The following example of a CFG expresses that NP (or noun phrase) can be composed of either a *ProperNoun* or a determiner (*Det*) followed by a *Nominal*; which can be one or more *Nouns*.

$$NP \rightarrow Det\ Nominal$$
$$NP \rightarrow ProperNoun$$
$$Nominal \rightarrow Noun \mid Nominal\ Noun$$

Legal Argument Constructs. The development of any grammar is factored by an initial linguistic analysis which determines the grammar symbols and the rules that allow to move from one symbol to another. The development of grammars of English focused on syntactic studies of individual sentences, while for legal arguments we must examine sentences in relationships.

In [41], ten documents from the ECHR collection were analysed by legal experts. The analysis broadly covered the section structure, argumentative structure, and linguistic characteristics found in the documents. Detailed analysis was done on those sections where the legal arguments were specifically presented – *The Law* and *Dissenting Opinion*. A clear distinction was observed between argumentative and non-argumentative information. The linguistic analysis identified some patterns only used on the argumentative information. For example, it clearly identified the conclusion of arguments, *"For these reasons the Court|Commission"*, which were supported by premises, *"There is a violation of Article"*. A premise of one argument can also serve as the conclusion of another argument, making chains fo arguments. The study also identified many rhetorical markers that help to detect the discursive progress of the argument structure, e.g. *however*, *therefore*, *although* or *in particular*. Furthermore, many premises and conclusions were found to be marked by linguistic expressions that clearly identified them as being part of the argumentative process (Table 1[6]). Some expressions may be common to all kind of argumentative texts, e.g. *"in the light of"*, while others may be more restricted to the legal field, e.g. *"under the terms of article"*.

Table 1. Typical expressions in the ECHR documents

Conclusions	The *factfinder* [A|B] <Verb-Conclusion> [C] The *factfinder* <Verb-Aux> [NOT] [A|B] <Verb-Conclusion> The *factfinder* [A|B] <Verb-Premise> There has been a violation of It [A] follows that There has [A] been a breach of Having reached this conclusion [C] In conclusion,
Premises	The *factfinder* [A|B] <Verb-Premise> [C] The *factfinder* <Verb-Aux> [NOT] [A|B] <Verb-Premise> [C] The *factfinder* [B] has|is [A] <Verb-Premise> [C] In the *factfinder*'s view In the view of the *factfinder* See, mutatis mutandis
[A]	therefore | firstly | accordingly | clearly | also | further | thus
[B]	like the xxxx and the xxxx, | , like the xxxx,
[C]	in the light of the partie's submissions | , in the light of all the material before it
<Verb-Conclusion>	accepts | concludes | holds | decides | rejects | declares | dismisses | sees no reason | examines | strikes
<Verb-Premise>	considers | notes | recalls | agrees | disagrees | reiterates | acknowledges | is of the opinion | points out | emphasises | stresses | is of the view | is satisfied | endorses | observes | takes into acount | convinces
<Verb-Aux>	must | can | does

A Context Free Grammar for Legal Argument. The grammar represents linguistic characteristics of legal argument with rules such as:

$$\forall_x [isPremise(x_i) \wedge startsHowever(x_{i+1}) \rightarrow isPremise(x_{i+1})]$$

To handle the linguistic variety required for real-world text input we compress families of related productions by making classes of rhetorical markers and verbs, e.g. support markers or conclusive verbs. We can generalise the previous rule as:

$$\forall_x [isPremise(x_i) \wedge startsContrast(x_{i+1}) \rightarrow isPremise(x_{i+1})]$$

---

[6] A *factfinder* is the person or persons in a particular trial or proceeding with the responsibility of determining the facts. For example, in the ECHR decisions the *factfinder* it is a *Commission*, in the ECHR judgments it is a *Court* and in other documents it is a *Judicial Responsible*, a *Committee* or a *jury*.

where $startsContrast(x_{i+1})$ is a class that contains $startsHowever(x_{i+1})$ among others. Thus, lexical and phrasal variation can be associated with similar semantics, allowing us to capture variations in writing style. The grammar does not accommodate ill-formed arguments as it would degrade the ability of the grammar to parse good arguments. The complete CFG can be found in Figure 1 and the meaning of the symbols in Table 2.

$$T \Rightarrow A^+ D \tag{1}$$

$$A \Rightarrow \{A^+ C | A^* CnP^+ | Cns | A^* src C | P^+\} \tag{2}$$

$$D \Rightarrow r_c f \{v_c s|.\}^+ \tag{3}$$

$$P \Rightarrow \{P_{verbP} | P_{art} | PP_{sup} | PP_{ag} | sP_{sup} | sP_{ag}\} \tag{4}$$

$$P_{verbP} = sv_p s \tag{5}$$

$$P_{art} = sr_{art} s \tag{6}$$

$$P_{sup} = \{r_s\}\{s | P_{verbP} | P_{art} | P_{sup} | P_{ag}\} \tag{7}$$

$$P_{ag} = \{r_a\}\{s | P_{verbP} | P_{art} | P_{sup} | P_{ag}\} \tag{8}$$

$$C = \{r_c | r_s\}\{s | C | P_{verbP}\} \tag{9}$$

$$C = s^* v_c s \tag{10}$$

Fig. 1. Context Free Grammar for recognising legal argument

Table 2. Terminal and non-terminal symbols from the Context Free Grammar

T	General argumentative structure of legal case
A	Argumentative structure that leads to a final decision of the factfinder $A = \{a_i, ..., a_j\}$, each a_i is an argument from the argumentative structure
D	The final decision of the factfinder $D = \{d_i, ..., d_j\}$, each d_i is a sentence of the final decision
P	One or more premises $P = \{p_i, ..., p_j\}$, each p_i is a sentence classified as premise
C	Sentence with a conclusive meaning
n	Sentence, clause or word that indicates one or more premises will follow
r_c	Conclusive rhetorical marker (e.g. therefore, thus, ...)
r_s	Support rhetorical marker (e.g. moreover, furthermore, also, ...)
r_a	Contrast rhetorical marker (e.g. however, although, ...)
r_{art}	Article reference (e.g. terms of article, art. x para. x, ...)
v_p	Verb related to a premise (e.g. note, recall, state, ...)
v_c	Verb related to a conclusion (e.g. reject, dismiss, declare, ...)
f	The entity providing the argument (e.g. court, jury, commission, ...)
s	Sentence, clause or word different from the above symbols

An Example of the Grammar's Application. The following example of an argument is taken from an ECHR judgment. We apply our CFG to it to generate a parse tree as in Figure 2.

The Court notes, firstly, that the applicant was convicted by the Greek courts of disturbing, through his writings, the public peace and the peace of the citizens of Western Thrace. Like the delegate of the Commission, the Court considers that the applicant's heirs also have a definite pecuniary interest under article of the convention art. x. Furthermore, it notes that the applicant was sentenced to fifteen months' imprisonment, commutable to a fine of x GRD per day of detention, which sum he paid. Without prejudice to its decision on the objection relating to non-exhaustion of domestic remedies, the Court considers that Mr. Ahmet Sadik's widow and children have a legitimate moral interest in obtaining a ruling that his conviction infringed the right to freedom of expression which he relied on before the convention institutions. The Court accordingly finds that Mrs. Isik Ahmet and her two children, Mr. Levent Ahmet and Miss. Funda Ahmet, have standing to continue the present proceedings in the applicant's stead.

The CFG identifies this as an argument with a conclusion, i.e. $A \Rightarrow A^+ C$. The conclusion, *"The Court accordingly finds that Mrs. Isik Ahmet and her two children, Mr. Levent Ahmet and Miss. Funda Ahmet, have standing to continue the present proceedings in the applicant's stead."*, is identified by:

– r_c: *accordingly*
– v_c: *finds*
– s: *that Mrs. Isik Ahmet and her two children, Mr. Levent Ahmet and Miss. Funda Ahmet, have standing to continue the present proceedings in the applicant's stead*

A^+ can be expanded as three separate premises using $A \Rightarrow P^+$. Each set of premises applies respectively:

– $P \Rightarrow P_{verbP}$ identifying v_p: *notes*
– $P \Rightarrow PP_{sup}$ and $P \Rightarrow P_{verbP}$ identifying r_s: *furthermore* and v_p: *considers*
– $P \Rightarrow P_{art}$ identifying $r_a$rt: *without prejudice to its decision on*

Thus, the final argument structure is the one found in Figure 2.

There are several interesting research issues concerning this grammar. First, it allows for some ambiguity. For example, the argument A, which has been treated with $A \Rightarrow P^+$ to obtain three P, being P_{verbP}, $PP_{sub}andP_{art}$ respectively, could have been divided in four P instead. Furthermore, the grammar relates a premise starting with a r_s or r_a to the closest previous statement, $P \Rightarrow \{P P_{sup} | P P_{ag}\}$, while that is not always the correct assumption. Sentences may not overtly express how they function. For example, the sentence *"Without prejudice to its decision on the objection relating to non-exhaustion of domestic remedies, the Court considers that Mr. Ahmet Sadik's widow and children have a legitimate moral interest in obtaining a ruling that his conviction infringed the right to freedom of expression which he relied on before the convention institutions."* seems to be a conclusion, but must be justified. If we cannot find support for such a justification, our grammar does not classify the statement as a conclusion. Legal arguments need justified conclusions for otherwise we cannot identify the argument. Finally, we can compare arguments allowing us to identify similar arguments written in different ways. For example, we can equate *"notes"* with *"acknowledges"*

```
A
|---c: The Court accordingly finds that Mrs. Isik Ahmet and her two
    |   children, Mr. Levent Ahmet and Miss Funda Ahmet, have standing
    |   to continue the present proceedings in the applicant's stead.
    |---A
        |---P
        |   |---p: The Court notes, firstly, that the applicant
        |   |       was convicted by the greek courts of disturbing,
        |   |       through his writings, the public peace and the
        |   |       peace of the citizens of western thrace.
        |---P
        |   |---p: Like the delegate of the Commission, the Court
        |   |       considers that the applicant's heirs also have a
        |   |       definite pecuniary interest under article of the
        |   |       convention art. x.
        |   |---P
        |       |---p: Furthermore, it notes that the applicant was
        |               sentenced to fifteen months' imprisonment,
        |               commutable to a fine of x GRD per day of
        |               detention, which sum he paid.
        |---P
            |---p: Without prejudice to its decision on the objection
                    relating to non-exhaustion of domestic remedies, the
                    Court considers that Mr. Ahmet Sadik's widow and
                    children have a legitimate moral interest in obtaining
                    a ruling that his conviction infringed the right to
                    freedom of expression which he relied on before the
                    convention institutions.
```

Fig. 2. Tree Structure of an argument

or *"that Mrs. Isik Ahmet and her two children, Mr. Levent Ahmet and Miss. Funda Ahmet, have standing to continue the present proceedings in the applicant's stead"* with *"that Mr. Smith is guilty"*. The grammar can be adapted to reflect the practices of the ECHR in this way. The comparison of arguments is crucial for applications of case based reasoning.

Grammar Evaluation. The grammar was tested to identify the argumentative structure of twenty new ECHR documents. It was implemented using Java and JSCC[7]. The main results can be seen in Table 3. It is important to note that all final decisions (D) were correctly identified and the average compression range of the given text was 65%. The main limitations of the grammar are due to the structure of A, i.e. the justification

Table 3. Results over 20 documents from the ECHR

	Size (# of sentences)	Precision	Recall
Premises	430	59%	70%
Conclusions	156	61%	75%
Non-argumentative information	1087	89%	80%
Final decision	63	100%	100%

---

[7] http://jscc.jmksf.coml

given by the *factfinder*. In this aspect, there are two main problems: (a) the detection of intermediate conclusions, specially the ones without rhetorical markers, as more than 20% of the conclusions are classified as premises of a higher layer conclusion; (b) the ambiguity between argument structures. However, this is also one of the main causes of human disagreement.

4.3   Wyner and Milward

Previous work such as [37] focuses on identifying argument structure from text such as indicated by keywords *supposing* or *therefore*, which are marks of argument in general and not clearly particular to legal argument. In legal argument in common law settings, there is a meta-level of argument concerning the cases themselves – case based reasoning (CBR) ([50], [51], [52], and [53]). CBR has four stages: the lawyer submits a problem case and retrieves precedent cases from a case base; the solutions from precedent cases are reused; the solution is confirmed (or disconfirmed); finally, once solved, the problem case is retained in the case base. From among these stages, a key task for a lawyer is to identify on-point-cases from a case base; these are cases which were decided in favour of the lawyer's side and share the most number of highly valued "factors". Factors are textually expressed typical fact patterns in a case which bias the decision for or against a side in the case. Importantly, factors are not themselves signalled by argumentative indicators. Yet, identifying the cases which contain the factors is crucial to case based reasoning. One may say that the second and third stages of CBR constitute the "argument" phase, where the argument is on analogy: given case A and case B, which are analogous to case C and which were decided for the plaintiff, therefore decide case C for plaintiff. [23] provide argument schemes for legal case based reasoning which detail the ways the factors are used to reason to a decision in a variety of instances.

For instance, suppose we are considering a case of reckless driving, where someone has died. Determining whether the driver is guilty of murder or manslaughter is crucial in the determination of the sentence. To make the distinction, it must be determined whether the driver had culpable intent. In turn, this is determined with respect to a variety of particular factors that can be concretely identified [54]:

- Obligation to aid the victim.
- Failure to heed traffic signs.
- Failure to heed warnings about reckless driving.

This list of factors which are used to determine culpable intent provide a textual "frame" for each factor; however, the form of the factors as they appear in a case base may vary. One of the important tasks of information extraction is to reliably identify the "same" factor in semantic terms while varying the form. This is a general linguistic issue (e.g. passives and actives mean the same thing, but have different forms).

We can also consider identifying factors across domains. For example, the issue relating to murder or manslaughter arises in the medical domain as well, and so culpable intent is also relevant. To determine the seriousness of medical negligence, one might consider:

- Obligation to have taken a second opinion.
- Failure to take a proper history.
- Failure to take into account apparent symptoms.

Thus, with one suitably general frame, one can identify cases addressing culpable intent across domains.

In searching the case base, one identifies the factors which contribute to the current undecided case, then wants to search the case base for on-point precedents. By the same token, one might want to identify cases which are *different* from the current case in order to compare the results in those cases to the current case.

In searching the case base, we applied a the commercial I2E text mining package by *Linguamatics*. The objective was to identify the factors as well as more general features of each case such as the citation, presiding judges, solicitors, whether the case is on appeal, and other parameters; these features are similar to the index parameters found in *Lexis-Nexis* and *Westlaw*, however, we identify these automatically rather than by preindexing the cases via manual annotation. I2E is an interactive, flexible, and articulated search tool; one specifies a search, views the results, then can refine or alter the search. It has a graphical user interface, which makes the software accessible to a broad range of end-users. The sentences in the case base are parsed, so searches can be done relative to syntactic structure. Additional search capabilities are: regular expression searches, list of alternative words, searches within syntactic frames such as sentences or paragraphs. A key feature is integration of searches relative to an ontology. For example, suppose one has a database of cases concerning medical malpractice and cancer. We may wish to relate doctors to the cancers they have treated. However, as there are a variety of cancers, using a string search alone, one would have to search relative to each sort of cancer expressed as a string. Given a suitably rich ontology, we can search just for the class "cancer", then retrieve all the different ways of referring to cancer (such as "carcinoma" or "neoplasm"), the different types of cancer, and the ways the types are referred to. In this way, we can relate parts of the text which are otherwise hard to relate with simple string searches.

In the following, we present several results.[8] In Table 4, we profile some cases from the case base: the document index (e.g. [2008]EWCACiv10.txt) shows a variety of features: the case number, the court in which the case is heard, who the solicitors were instructed by, the judge who hears the case, and the court from which the case is appealed.

In Table 5, we identify the concept of *Failure/Obligations* from a case in the case base; the concept appears in a variety of alternative phrases: *ought not to have...*, *had failed to observe...*, *owed a duty of care....* This highlights phrases which can be used to further determine whether the case concerns medical malpractice.

For tasks where the same kind of information needs to be automatically extracted, the next step is to develop a corpus of cases as a gold standard against which to measure information retrieval with respect to precision and recall. However, there are also specific tasks where it is useful to apply the graphical querying capabilities of I2E which are similar to ad hoc keyword search, and iteratively refine queries to get the cases of interest.

---

[8] These are simplified results produced by I2E.

Table 4. Case Profile

Doc	Index	Entity
[2008] EWCACiv10.txt	Case No.	A3/2007/1677
	Court	The Supreme Court of Judicature Court of Appeal
	Instructed by	Skadden
		Steptoe & Johnson
	On appeal	High Court of Justice Queen's Bench Division Commerical Court
[2008] EWCACiv1022.txt	Case No.	B2/2007/2303
	Court	The Supreme Court of Judicature Court of Appeal
	Instructed by	Messrs Buller Jeffries
		Messrs John A Neil Solicitors
	Judge	District Judge Temple
	On appeal	Cambridge County Court

Table 5. Case Profile with Violation Factor

Doc	Index	Entity	Context
30.htm	Case No.	200302858 B1	
	Court	The Supreme Court of Judicature	
	Failure or Obligations	had failed to observe appropriate professional standards	in the sense that he had failed to observe apprporiate professional standards to a patient to whom...
		owed a duty of care	as a doctor he owed a duty of care to Sean Philips as his...
		owes a duty of care	individual to whom the defendant owes a duty of care
	Judge	Lord Justice Judge	
	On appeal	Winchester Crown Court	

5   Discussion

We have set our work in the context of argumentation theory and text mining in the case base. We have discussed different approaches to the formation of a legal case base. Finally, we turned to several ways elements found in a case base can be automatically analysed with text mining tools to support legal argument. In this section, we consider issues with our results as well as a range of topics that warrant future research.

With regard to the development of legal argument corpora, there are new opportunities to use available online corpora for text analysis as noted in section 3.4. However, some consideration should be given to the signficance of using different corpora given the great variety of court systems. For example, [41] use decisions from the European Court of Human Rights. In related work on diagramming legal decisions, [55] use US Supreme Court oral argument transcripts. In both instances, consideration should be given to the level of court and complexity of issues since both court levels address areas of the law that are the least settled and most complex. Examining US Supreme Court transcipts, the argument patterns are very complex and often hard to follow, even by the Justices and legal representative themselves as is reported in court transcripts. It may be preferable to make use of court levels and decisions that are signicantly more prosaic such as decisions of courts in of the first instance (where the case is first introduced) or courts of appeals. Not only is the law more settled and less complex, but it is more useful to legal researchers to know how the law commonly functions before we examine how it functions at the highest level (on this point, also see [56]).

In section 4.2, a grammar of argument is provided and then used to extract arguments from cases. Such an approach may work well for well-edited court decisions, but it is unlikely to apply to oral arguments made in court by lawyers, which appear less structured[9] It is, then, important to proscribe just what is intended to be meant by a grammar of *legal* arguments. Moreover, even with well-edited court decisions, it is as yet unclear the extent to which *discontinuous constituents* (see [57]) or discourse phenomena (see [58]) play a role. The proposal in section 4.2 relies on continuous structures such as for phrase-structure grammars for sentences. In such an approach, phrases must appear in determined orders, must be complete, cannot allow components of the argument appearing outside the given structure, and do not allow interjections. Yet, it is possible that the premise of an argument appears somewhere later in the text, which a CFG alone could not account for. However, in the corpus studied, it appears that the percentage of discontinuous constituents is normally low. Nonetheless, some consideration ought to go into how to accommodate discontinuous constituents where they occur.

Another aspect of the grammar of argument is that only *defeasible* arguments are accounted for, not the variety of argument types outlined in [5]. To identify subsorts of argument (e.g. Expert Testimony), lexical semantic and ontological information may be required.

Other issues of interest relate to the identification of enthymemes, which are missing and inferred premises, as well as argument coherence. For information extraction, enthymemes present a very significant issue since they must be inferred semantically and are not in evidence in the text. It is unclear how text mining can yet address this issue. In section 4.2, the pattern of an argument is identified, but within the components of the argument (i.e. premises, rule, conclusion) there appear to be no well-formedness constraints. Yet, clearly, there are well-formed arguments in terms of the grammar, but which are semantically incoherent: if Tweety is a bird and iron is a mineral, therefore the stock market will rise. This relates to similar syntactic issues bearing on sentences such as the famous *Colorless green ideas sleep furiously*. Of course, the research reported in section 4.2 is a step towards identifying, clarifying, and addressing such issues.

In section 4.3, initial results are provided for automatically profiling a case and identifying key features that are useful in making legal arguments. Further research will focus on identifying legally relevant case factors such as are actually used by lawyers in case based reasoning. Here too, lexical semantic and ontological information may prove to be useful.

---

[9] See US Supreme Court transcripts:
http://www.supremecourtus.gov/oral_arguments/argument_transcripts.html

A general issue that must be addressed is the bottleneck of ontological approaches, which still require large amounts of knowledge to be manually drafted. Such a task is not realistic for general text analysis. One way to address the issue is to apply advanced machine-learning techniques. Another is to provide a structure that supports systematic, incremental, modular knowledge development such as is done with Semantic Web OWL modules. Yet another approach is to leverage the internet to distribute the task of manually drafting knowledge by, for example, using techniques such as found for online psycholinguistic experiments.[10]

The results described in this paper bring us closer to automatic identification of legal arguments. We believe that the use of ontologies, which were used in profiling, will also be useful for identifying arguments. In future work we will look at combining the two approaches examined here to provide a single approach for comprehensive analysis of legal texts.

Acknowledgments

[41] was presented at JURIX 2008. [42] was presented at the Workshop on Natural Language Engineering of Legal Argumentation held in conjunction with JURIX 2008. The authors thank the reviewers and audiences for their comments.

References

1. Wyner, A., Bench-Capon, T., Atkinson, K.: Three senses of "argument". In: Casanovas, P., Sartor, G., Casellas, N., Rubino, R. (eds.) Computable Models of the Law. LNCS (LNAI), vol. 4884, pp. 146–161. Springer, Heidelberg (2008)
2. Eemeren, F.V., Grootendorst, R.: A Systematic Theory of Argumentation. Cambridge University Press, Cambridge (2003)
3. Mann, W.C., Thompson, S.A.: Rhetorical structure theory: A theory of text organization. Technical Report ISI/RS-87-190, University of Southern California, Information Sciences Institute, ISI (1987)
4. Hitchcock, D.: Informal logic and the concept of argument. In: Gabbay, D., Thagard, P., Woods, J. (eds.) Philosophy of Logic (Handbook of the Philosophy of Science), pp. 101–130. Elsevier, Amsterdam (2007)
5. Walton, D.: Argumentation Schemes for Presumptive Reasoning. Erlbaum, Mahwah (1996)
6. Atkinson, K.: What Should We Do? Computational Representation of Persuasive Argument in Practical Reasoning. PhD thesis, Department of Computer Science, University of Liverpool, Liverpool, United Kingdom (2005)
7. Gordon, T., Prakken, H., Walton, D.: The carneades model of argument and burden of proof. Artificial Intelligence 171, 875–896 (2007)
8. Reed, C., Rowe, G.: Araucaria: Software for argument analysis, diagramming and representation. International Journal on Artificial Intelligence Tools 13(4), 961–980 (2004)
9. Torroni, P., Gavanelli, M., Chesani, F.: Argumentation in the semantic web. In: Rahwan, I., McBurney, P. (eds.) Intelligent Systems, vol. 22(6), pp. 66–74 (2007)
10. Dung, P.M.: On the acceptability of arguments and its fundamental role in nonmonotonic reasoning, logic programming and n-person games. Artificial Intelligence 77(2), 321–358 (1995)

---

[10] See: http://www.surf.to/experiments

11. Amgoud, L., Cayrol, C.: On the acceptability of arguments in preference-based argumentation. In: Proceedings of the 14th Annual Conference on Uncertainty in Artificial Intelligence (UAI 1998), pp. 1–7. Morgan Kaufmann, San Francisco (1998)

12. Bench-Capon, T.J.M.: Persuasion in practical argument using value-based argumentation frameworks. Journal of Logic and Computation 13(3), 429–448 (2003)

13. Besnard, P., Hunter, A.: Elements of Argumentation. MIT Press, Cambridge (2008)

14. Wyner, A., Bench-Capon, T.: Towards an extensible argumentation system. In: Mellouli, K. (ed.) ECSQARU 2007. LNCS (LNAI), vol. 4724, pp. 283–294. Springer, Heidelberg (2007)

15. Bondarenko, A., Dung, P.M., Kowalski, R.A., Toni, F.: An abstract, argumentation-theoretic approach to default reasoning. Artificial Intelligence 93, 63–101 (1997)

16. Wyner, A., Bench-Capon, T.: Taking the a-chain: Strict and defeasible implication in argumentation frameworks. Technical report, University of Liverpool (2008)

17. Prakken, H., Sartor, G.: Argument-based extended logic programming with defeasible priorities. Journal of Applied Non-Classical Logics 7(1) (1997)

18. García, A.J., Simari, G.R.: Defeasible logic programming: An argumentative approach. Theory and Practice of Logic Programming 4(1), 95–137 (2004)

19. Governatori, G., Maher, M.J., Antoniu, G., Billington, D.: Argumentation semantics for defeasible logic. Journal of Logic and Computation 14(5), 675–702 (2004)

20. Amgoud, L., Caminada, M., Cayrol, C., Lagasquie, M.C., Prakken, H.: Towards a consensual formal model: inference part. Technical report, ASPIC project, Deliverable D2.2: Draft Formal Semantics for Inference and Decision-Making (2004)

21. Pollock, J.: Cognitive Carpentry: A Blueprint for How to Build a Person. MIT Press, Cambridge (1995)

22. Caminada, M., Amgoud, L.: On the evaluation of argumentation formalisms. Artificial Intelligence 171(5-6), 286–310 (2007)

23. Wyner, A., Bench-Capon, T.: Argument schemes for legal case-based reasoning. In: Lodder, A.R., Mommers, L. (eds.) Legal Knowledge and Information Systems, JURIX 2007, pp. 139–149. IOS Press, Amsterdam (2007)

24. Bos, J.: Applying automated deduction to natural language understanding. Journal of Applied Logic 7(1), 100–112 (2008)

25. Curran, J.R., Clark, S., Bos, J.: Linguistically motivated large-scale NLP with C&C and Boxer. In: ACL, The Association for Computer Linguistics (2007)

26. MacCartney, B., Manning, C.D.: Natural logic for textual inference. In: Proceedings of the ACL-PASCAL Workshop on Textual Entailment and Paraphrasing, Prague, Association for Computational Linguistics, pp. 193–200 (2007)

27. de Paiva, V., Bobrow, D.G., Condoravdi, C., Crouch, D., King, T.H., Karttunen, L., Nairn, R., Zaenen, A.: Textual inference logic: Take two. In: C&O:RR (2007)

28. Dagan, I., Glickman, O., Magnini, B.: The PASCAL Recognising Textual Entailment Challenge. In: Quiñonero-Candela, J., Dagan, I., Magnini, B., d'Alché-Buc, F. (eds.) MLCW 2005. LNCS (LNAI), vol. 3944, pp. 177–190. Springer, Heidelberg (2006)

29. de Marneffe, M.C., Rafferty, A.N., Manning, C.D.: Finding contradictions in text. In: Proceedings of ACL 2008: HLT, Columbus, Ohio, Association for Computational Linguistics, pp. 1039–1047 (2008)

30. Voorhees, E.M.: Contradictions and justifications: Extensions to the textual entailment task. In: Proceedings of ACL 2008: HLT, Columbus, Ohio, Association for Computational Linguistics, pp. 63–71 (2008)

31. Hafner, C.D.: Representation of knowledge in a legal information retrieval system. In: SIGIR 1980: Proceedings of the 3rd annual ACM conference on Research and development in information retrieval, Kent, UK, pp. 139–153. Butterworth & Co. (1981)

32. Daniels, J.J., Rissland, E.L.: Finding legally relevant passages in case opinions. In: ICAIL 1997: Proceedings of the 6th International Conference on Artificial intelligence and Law, pp. 39–46. ACM, New York (1997)

33. Brüninghaus, S., Ashley, K.D.: Finding factors: learning to classify case opinions under abstract fact categories. In: ICAIL 1997: Proceedings of the 6th International Conference on Artificial Intelligence and Law, pp. 123–131. ACM, New York (1997)

34. Brüninghaus, S., Ashley, K.D.: Generating legal arguments and predictions from case texts. In: ICAIL 2005, pp. 65–74. ACM Press, New York (2005)

35. Jackson, P., Al-kofahi, K., Tyrell, A., Vachher, A.: Information extraction from case law and retrieval of prior cases. Artificial Intelligence 150(1-2), 239–290 (2003)

36. McCarty, L.T.: Deep semantic interpretations of legal texts. In: ICAIL 2007: Proceedings of the 11th International Conference on Artificial Intelligence and Law, pp. 217–224. ACM Press, New York (2007)

37. Moens, M.F., Boiy, E., Mochales-Palau, R., Reed, C.: Automatic detection of arguments in legal texts. In: ICAIL 2007: Proceedings of the 11th International Conference on Artificial Intelligence and Law, pp. 225–230. ACM Press, New York (2007)

38. Sporleder, C., Lascarides, A.: Using automatically labelled examples to classify rhetorical relations: An assessment. Natural Language Engineering 14(3), 369–416 (2006)

39. Schweighofer, E.: The revolution in legal information retrieval or: The empire strikes bac. The Journal of Information, Law and Technology 1 (1999)

40. Reed, C., Palau, R.M.P., Rowe, G., Moens, M.F.: Language resources for studying argument. In: Proceedings of the 6th conference on language resources and evaluation - LREC 2008. ELRA, pp. 91–100 (2008)

41. Mochales, R., Moens, M.F.: Study on the structure of argumentation in case law. In: Francesconi, E., Sartor, G., Tiscornia, D. (eds.) Legal Knowledge and Information Systems - JURIX 2008: The Twenty-First Annual Conference, Frontiers in Artificial Intelligence and Applications, vol. 189, pp. 11–20. IOS Press, Amsterdam (2008)

42. Wyner, A., Milward, D.: Legal text-mining using linguamatics' I2E. Presentation at Workshop on Natural Language Engineering of Legal Argumentation, Florence, Italy as part of JURIX 2008 (2008)

43. Chesnevar, C., McGinnis, J., Modgil, S., Rahwan, I., Reed, C., Simari, G., South, M., Vreeswijk, G., Willmott, S.: Towards an argument interchange format. The Knowledge Engineering Review 21(4), 293–316 (2006)

44. Jackendoff, R.: Semantic Interpretation in Generative Grammar. MIT Press, Cambridge (1972)

45. Walton, D.: The three bases for the enthymeme: A dialogical theory. Journal of Applied Logic 6(3), 361–379 (2008)

46. Biber, D., Johansson, S., Leech, G., Conrad, S., Finegan, E. (eds.): The Longman Grammar of Spoken and Written English. Longmans, London (1999)

47. Feldmann, H.: An acceptive grammar for the natural language english. SIGPLAN Not. 19(2), 58–67 (1984)

48. Scott, W.: Development and application of a context-free grammar for requirements. In: International Conferences and Events (ICE), Australia (2004)

49. Charniak, E.: Statistical parsing with a context-free grammar and word statistics. AAAI Press/MIT Press (1997)

50. Ashley, K.: Modelling Legal Argument: Reasoning with Cases and Hypotheticals. Bradford Books/MIT Press (1990)

51. Aleven, V., Ashley, K.D.: Doing things with factors. In: ICAIL 1995: Proceedings of the 5th International Conference on Artificial Intelligence and Law, pp. 31–41. ACM, New York (1995)

52. Weber, R.O., Ashley, K.D., Brüninghaus, S.: Textual case-based reasoning. Knowledge Engineering Review 20(3), 255–260 (2005)
53. Brüninghaus, S., Ashley, K.: Reasoning with textual cases. In: Muñoz-Ávila, H., Ricci, F. (eds.) ICCBR 2005. LNCS (LNAI), vol. 3620, pp. 137–151. Springer, Heidelberg (2005)
54. Luria, D.: Death on the highway: Reckless driving as murder. Oregon Law Review 799, 821–822 (1988)
55. Ashley, K.D., Pinkwart, N., Lynch, C., Aleven, V.: Learning by diagramming supreme court oral arguments. In: ICAIL, pp. 271–275. ACM, New York (2007)
56. Prakken, H.: Formalising ordinary legal disputes: a case study. Artificial Intelligence and Law 16(4), 333–359 (2008)
57. Ojeda, A.: Discontinuous Constituents. In: Encyclopedia of Language and Linguistics, vol. 3, pp. 624–630. Elsevier, Amsterdam (2005)
58. Asher, N.: Reference to Abstract Objects in Discourse. Kluwer Academic Publishers, Dordrecht (1993)

PART II

Legal Text Processing and Construction of Knowledge Resources

Automatic Identification of Legal Terms
in Czech Law Texts

Karel Pala, Pavel Rychlý, and Pavel Šmerk

Faculty of Informatics
Masaryk University
Botanická 68a, 602 00 Brno
Czech Republic
{pala,pary,smerk}@mail.muni.cz

Abstract. Law texts including constitution, acts, public notices and court judgements form a huge database of texts. As many texts from small domains, the used sublanguage is partially restricted and also different from general language (Czech). As a starting collection of data, the legal database Lexis containing approx. 50,000 Czech law documents has been chosen. Our attention is concentrated mostly on noun groups, which are the main candidates for law terms. We were able to recognize 3992 such different noun groups in the selected text samples. The paper also presents results of the morphological analysis, lemmatization, tagging, disambiguation, and the basic syntactic analysis of Czech law texts as these tasks are crucial for any further sophisticated natural language processing. The verbs in legal texts have been explored preliminarily as well. In this respect, we are trying to explore how the linguistic analysis can help in identification of the semantic nature of law terms.

Keywords: Terminology Extraction, Natural Language Processing, Legal Language.

1 Introduction

In the paper we describe the first results of the new project whose final goal is to build an electronic dictionary of Czech law terms. In this task we cooperate with the teams from the Institute of Government and Law and the Institute of Czech Language, Czech Academy of Sciences in Prague. We have started with a legal database Lexis developed at the Institute of Law, which presently includes approx. 50,000 Czech law documents ranging from the beginning of Czechoslovak Republic in 1918 to the present day. It also includes court judgements, main representative law textbooks and law reports. All of these texts exist in electronic form.

The first part of the paper presents results of the preparation step for the subsequent term identification – the morphological analysis. For this purpose we have used the tools developed in the Natural Language Processing Centre of the Faculty of Informatics, Masaryk University, particularly, the morphological

E. Francesconi et al. (Eds.): Semantic Processing of Legal Texts, LNAI 6036, pp. 83–94, 2010.
© Springer-Verlag Berlin Heidelberg 2010

analyser **ajka** [1] performing lemmatization and tagging and a new tool for grammatical disambiguation named **desamb** [2]. The tools have been designed for general Czech but it appears that they can be utilized for the law sublanguage with some minor modifications, namely adding law terms. The tools are now configured to analyze all Czech law texts contained in the Lexis database.

In the second part, we report about term identification via syntactic analysis which has used the tool DIS/VADIS [3], a partial parser for Czech. As a result, list of noun groups has been obtained that can be considered as good candidates for law terms. We are also taking a look at the verbs existing in law texts because they are relational elements linking together the established law terms. Here the apparatus of valency frames [4] comes as an appropriate instrument. It allows us to explore context patterns in which law terms occur and see how they behave in the law text.

The general goal is to find out to what extent linguistic analysis can contribute to semantic analysis of the law text. We are at the starting point of this enterprise.

1.1   Pilot Project

As a pilot project we have decided to analyse the current version of the Penal Code of the Czech Republic. It is one of the biggest law documents containing almost 36,000 word forms. The overall characteristic of the document can be found in Table 1.

Table 1. The overall characteristic of the Penal Code of the Czech Republic

Number of	
word forms (tokens starting with a letter)	35,898
numbers (tokens containing digits)	2,832
punctuation marks (anything else)	9,135
tokens total	47,865
different word forms	4,400
different numbers	607
different punctuation marks	12
types total	5,019

The task is to process the document by the Czech morphological analyser (lemmatizer) **ajka** in such a way, that for each word form in the source text morphological information in the form of morphological tags is obtained. Thus we get information to what parts of speech the word forms belong, and, for instance, for nouns also grammatical categories like gender, number and case. Each word form in the document is associated with its respective lemma as well. In a highly inflectional language like Czech, all this information is relevant for the further analysis of law terms. The results of the morphological analysis and lemmatization were transformed into a special format which is described below.

2   Morphological Analysis

We have used several simple scripts to create a what is called "vertical" file from the source text. It is a plain text file without any formatting (word-processing options). Words are written in a column, i. e. each line contains one word, number or punctuation mark. Optional annotation is on the same line and the respective words are divided by the tabulator character. The first step uses only word forms from the source text. The vertical file serves as a standard input text for many corpus processing tools like CQP [5] and Manatee [6].

In the next step, the vertical file was processed with the morphological analyser ajka [1]. It is a tool used for annotating and lemmatizing general Czech texts, however, the processing of law texts requires modifications, e. g. enriching the list of stems of ajka. The program yields all possible combinations of lemma and morphological tags for each Czech word form.

Table 2 presents an example of the ajka output. E. g. the tag **k1gFnSc1** means: part of speech (**k**) = noun (**1**), gender (**g**) = female (**F**), number (**n**) = singular (**S**) and case (**c**) = first (nominative)(**1**). Tags beginning with **k2** are adjectives, **k3** are pronouns, **k5** verbs and **k7** prepositions.

Table 2. Output of the morphological analyser ajka

word	possible lemmata (<l>) and tags (<c>)
Příprava (preparation)	<l>příprava <c>k1gFnSc1
k (to)	<l>k <c>k7c3
trestnému (criminal)	<l>trestný <c>k2eAgMnSc3d1 <c>k2eAgInSc3d1 <c>k2eAgNnSc3d1
činu (act)	<l>čin <c>k1gInSc3 <c>k1gInSc6 <c>k1gInSc2 <l>čina <c>k1gFnSc4
je (is)	<l>být <c>k5eAaImIp3nSrDaI <l>on <c>k3p3gMnPc4xP <c>k3p3gInPc4xP <c>k3p3gNnSc4xP <c>k3p3gNnPc4xP <c>k3p3gFnPc4xP <l>je <c>k0
trestná (criminal)	<l>trestný <c>k2eAgFnSc1d1 <c>k2eAgFnSc5d1 <c>k2eAgNnPc1d1 <c>k2eAgNnPc4d1 <c>k2eAgNnPc5d1

As one can see, many word forms are ambiguous: there are more than one possible tag or even lemma for a given word form. In the analysed document, 76 % of word forms are ambiguous, more than 42 % of word forms have more than one possible lemma and average number of tags for an ambiguous word form is 6.75.

We have used part-of-speech tagger desamb with the success rate 95.15 % [2] to disambiguate such word forms. The output of the desamb tool contains only the most probable lemma and tag for each word form. Table 3 contains output of desamb for the input text above.

Table 3. The document in vertical format with morphological annotation (after disambiguation)

word	lemma	tag
Příprava	příprava	k1gFnSc1
k	k	k7c3
trestnému	trestný	k2eAgInSc3d1
činu	čin	k1gInSc3
je	být	k5eAaImIp3nS
trestná	trestný	k2eAgFnSc1d1
podle	podle	k7c2
trestní	trestní	k2eAgFnSc2d1
sazby	sazba	k1gFnSc2
stanovené	stanovený	k2eAgFnSc2d1
na	na	k7c4
trestný	trestný	k2eAgInSc4d1
čin	čin	k1gInSc4

The annotated version of the document contains 2,560 different lemmas. Frequencies of each part of speech are in Table 4. The 'other' category covers abbreviations and paragraph letters.

Table 4. Frequencies of parts-of-speech in the document

POS	Count
k1 – noun	12,889
k2 – adjective	4,634
k3 – pronoun	2,252
k4 – numeral	1,028
k5 – verb	4,504
k6 – adverb	933
k7 – preposition	3,600
k8 – conjunction	3,764
k9 – particle	632
other	1,662

3   Noun Groups

For the recognition of the noun groups we have used the partial syntactic analyzer for Czech DIS/VADIS [3] at first. Unfortunately, DIS/VADIS presently does not contain rules which can recognize genitival and coordinate structures because during the development of DIS/VADIS, these rules were found too erroneous (overgenerating) when applied to unrestricted text. However, there are plenty of such structures in the law texts and overgenerating is not a problem here because the results will be checked manually.

Moreover, the partial syntactic analyzer DIS/VADIS has one more disadvantage: it is written in Prolog which implies that the recognition process is rather slow. Therefore we have rewritten the rules for noun groups to Perl 5.10 regular expressions (which have nontrivial backtracking capabilities) and added the rules for genitival and coordinate structures and some adverbials common to the law texts which also were not recognized by DIS/VADIS (e. g. *zvlášť* (exceedingly), *zjevně* (evidently) etc.).

For each noun group found in the law texts we determine its:

1. base form (nominative singular),
2. head,
3. for nouns in genitive groups also their part.

For example, for the noun group *dalším pácháním trestné činnosti* (subsequent commission of criminal activity, dative) we get:

1. *další páchání trestné činnosti,*
2. *páchání,*
3. *další páchání.*

We can recognize 8,594 noun groups counting repeating occurrences, 3,992 different noun groups. The noun groups were analyzed and the respective 'base' of each noun group was derived. Due to the inflectional feature of Czech, this cannot be done by simple lemmatization of all words in a noun group. The automatic transformation algorithm works in following steps:

− find dependences between parts (words of subgroups) of a noun group,
− locate the head − key word,
− identify matching noun group pattern,
− generate the correct word forms with matching grammatical categories.

The result of this algorithm are base forms of noun groups and they will appear as headwords in the final electronic dictionary. The most frequent base forms with respective number of occurrences in the pilot data are listed in Table 5. (There are some conceptual problems with finding the correct English equivalent terms thus we do not offer them here.)

Table 6 presents the most frequent part-of-speech patterns of the recognized noun groups. There are two counts in the table: 'Tokens' is the total number of occurrences of the respective pattern in the pilot data and 'Types' is the number of different noun groups matching such pattern. The tag k1 denotes a noun, k2 is an adjective, and g is a gender; in particular, gM stands for masculine animate, gI for masculine inanimate, gF for feminine, and gN for neuter.

Table 5. The most frequent terms (base forms)

Term	Count
odnětí svobody	568
trestný čin	228
peněžitý trest	152
jeden rok	123
zákaz činnosti	81
trest odnětí svobody	69
účinnost dne	65
(jiná) majetková hodnota	65
velký rozsah	64
těžká újma	58
výjimečný trest	51
organizovaná skupina	49
závažný následek	47
zvlášť závažný následek	46
veřejný činitel	46
značný prospěch	40
jiný zvlášť závažný následek	40
značná škoda	39
člen organizované skupiny	39
stav ohrožení státu	37

Table 6. The most frequent POS patterns

POS pattern	Tokens	Types
k2 – k1gI	1588	344
k2 – k1gF	1130	365
k1gN – k1gF	765	96
k2 – k1gN	478	213
k1gI – k1gN	204	57
k1gN – k1gI	203	80
k1gI – k1gF	195	67
k2 – k1gM	176	71
k2 – k2 – k1gF	163	65
k1gF – k1gI	162	48

4   Verbs and Verb Groups

Though law terms typically consist of the nouns, noun groups and other nominal constructions, we also have paid attention to the verbs found in the whole database of the 50,000 law documents. The reason for this comes from the fact, that verbs on one hand do not always display strictly terminological nature, but on the other hand they are relational elements linking the terminological nouns and noun groups together.

The verbs were originally processed by the team of F. Cvrček in the Institute of Government and Law. From them we received the list 15110 items marked as verbs. Then we used textttajka for further processing with the following results: 4,920 items in the list were marked as passive participles - they were not further lemmatized. After manual checking we discovered that 1611 items from them were not recognized as verbs but for example as adjectives or nouns and they were removed from the list. Thus the list of the correctly recognized verb lemmata comprises 10,190 items. Then we looked at the error list and observed shows that at least three types of the non-recognized items can be found:

1. erroneous forms caused by typing errors. They can be corrected, e. g. *citit* instead of the correct *cítit* (feel);
2. the verbs that do not appear in the textttajka's list of stems. Typically, they display terminological nature and they should be added to the **ajka**'s stem list, e. g. *derogovat* (derogate). In this way, they will enrich the list of (Czech) verb stems. Their law meanings constitute a terminological subset of verbs;
3. erroneous forms that cannot be corrected without correcting the whole paragraph of a law document (we have not touched them). Typically, they are incorrect lexical items in the particular paragraph they occur in. The correction would require to replace them with more precise ones, however, that would mean the change of the the whole law document.

The non-recognized verbs were added to **ajka**'s list of verb stems. The next step was to make an intersection of the resulting list of 8,579 'legal' verbs with our lexical database VerbaLex [4] containing presently 8,366 (general) Czech verbs. The VerbaLex list contains altogether 10,472 verbs including reflexive variants with the reflexive particles *se* and *si* (*vzít (to take), vzít se (to marry), vzít si (to take yourself)*).

The comparison offered the following result:

− 3,749 verbs occurring only in the legal texts,
− 3,563 verbs occurring only in VerbaLex,
− 4,830 verbs occurring in the both resources.

The obtained numbers show some tendencies that are in agreement with our expectations i. e. that many verbs in legal text have a clearly terminological character. It is typical for the verbs occurring in the intersection that they appear in the different meanings, i. e. in general and legal. At this point, more detailed analysis of the contexts in which these verbs appear is needed. This analysis is planned for the near future.

To get some basic information about the behaviour of the verbs in the three mentioned lists, we found the frequencies of the verbs in the Czech corpus SYN2000 [7]. For illustration, we offer ten to twelve most frequent verbs from each list in Tables 7, 8 and 9:

Table 7. 'Legal' verbs

slyšet (hear)	19526
smět (may, modal)	17939
hodlat (intend)	13415
narodit (be born)	8386
pravit (say)	7553
nosit (carry)	6063
zavolat (call)	5561
pozvat (invite)	5404
vyžádat (require)	4619
vstát (stand up)	4533
obejít (get around)	4516
hlasovat (vote)	3929

Table 8. Only VerbaLex verbs

moci (can, modal)	329852
říci (say, tell)	155508
pomoci (help)	20621
vyprávět (narrate)	7470
utéci (run away)	4481
kout (pikle) (plot)	2638
zasmát (laugh)	2309
téci (flow)	2253
vstřelit (score)	1818
linout (waft)	1652

Table 9. Intersection: both legal and VerbaLex verbs

být (be)	3388353
mít (have)	634214
muset (must, have to)	165621
chtít (want)	143495
jít (go)	123017
vědět (know)	94412
dát (give)	90221
začít (start)	74910
dostat (get)	69961
říkat (say, tell)	68078
vidět (see)	61979

The first ten verbs do not allow us to make more concrete conclusions since they are general verbs that occur in any texts. However, the less frequent verbs from the list of legal verbs on the other hand display specialized terminological meanings; for instance, following compound verbs do not occur in the corpus SYN2000 at all: *spoluvinit* (co-accuse), *spoluvázat* (co-bind), *spoluzabezpečovat* (co-ensure), *spoluzaviňovat* (co-cause), *spoluzavazovat* (co-oblige), *spoluzpůsobovat* (co-cause), *spoluzpůsobit* (co-cause, aspect counterpart of the previous one), *spolužalovat* (co-sue), etc.

It has been observed (unpublished report of the F. Cvrček and his team), that the legal verbs co-occur with the nouns which can be semantically included into the following groups:

1. one word and multi-word with the autonomous legal meaning, e. g. agreement or contract,
2. nouns with possible legal meaning that follow from the context in which it is used, e. g. person,
3. nouns with clearly non-legal meaning, e. g. chloride,
4. nouns that denote subjects or agents, for instance:
   (a) legal subject such as malefactor,
   (b) legal subject following from context, e. g. member,
   (c) employment, e. g. sculptor,
   (d) legally preferred group, e. g. pensioner,
   (e) subjects by nationality and race, e. g. Serbian, white man,
   (f) nouns with emotional and ideological connotation, such as angel, whore,
   (g) nouns denoting animals, e. g. whale, squirrel, etc.

The above mentioned semantic categories can be reasonably compared with the semantic roles as they are used in the verb valency frames from VerbaLex database, and it can be concluded that the valency frames are suitable for the semantic description of the legal language. We can observe the interesting overlaps which allow us to speak about the existence of semantic roles in legal texts. They can be easily added to the present inventory of the semantic roles in VerbaLex. This is a positive result which confirms the assumption that though legal language displays some specific features it can be analysed with techniques and methods developed for semantic analysis of verb meanings as they can be found in a non-terminological use.

In VerbaLex we work with the roles such as AG<person:1> etc., which in the legal language correspond to the 'subject' mentioned above. Thus it is possible to take advantage of the roles introduced in VerbaLex and apply them to the semantic categories mentioned above. In this way, for instance, we can easily obtain labels such as AG<judge:1>, AG<employee:1> or PAT<person:1> and other similar ones.

For some of the verbs occurring in the intersection list, VerbaLex already contains valency frames that capture their legal meanings. For instance, one of the frames of the verb *zabít* (kill) is

$$\text{AG<person:1>}_{obl}^{kdo1} \text{ VERB PAT<person:1>}_{obl}^{koho4} \text{ INS<instrument:1>}_{opt}^{cim7},$$

and similarly one of the frames of the verb *potrestat* (penalize) is

$$\text{AG<person:1>}_{obl}^{kdo1} \text{ VERB PAT<person:1>}_{obl}^{koho4} \text{ EVENT<punishment:1>}_{obl}^{cim7}.$$

Of course, there are many legal verbs that are not contained in VerbaLex. However, we can apply the notation developed for VerbaLex to write valency frames for them. Below, we adduce two more examples of the frames proposed for selected fully legal verbs. It is obvious can that the VerbaLex notation can serve appropriately for this purpose:

uložit trest někomu (to condemn somebody to a sentence)

$$\text{AG<judge:1>}_{obl}^{kdo1} \text{ VERB PAT<person:1>}_{obl}^{komu3} \text{ ACT<sentence:1>}_{obl}^{co4},$$

obvinit někoho z trestného činu (to accuse somebody of criminal act)

$$\text{AG<}_{\text{prosecutor}}^{\text{public}}\text{:1>}_{obl}^{kdo1} \text{ VERB PAT<person:1>}_{obl}^{koho4} \text{ ACT<act:1>}_{obl}^{zceho2}.$$

Labels used for the roles consist of two parts where the first one (e. g. AG as agent, PAT as patient etc.) was taken from the EuroWordNet Top Ontology (TO) and the second one comes from the set of the Base Concepts (BCs) also defined in EuroWordNet [8]. The Base Concepts are used as subcategorization features and they are represented formally by the Princeton WordNet literals together with their sense numbers as they exist within the individual synsets (e.g. person:1, instrument:1 etc.). It is important to realize that they also represent nodes in the hypero/hyponymy trees in (Princeton) WordNet [9], i. e. they are endogenous and allow us to describe the meanings of the verb arguments in a detailed way. In the frames we use indices – the upper ones following the roles express the surface valences, i. e. morphological cases that have to be indicated in Czech. The lower indices say that the particular role is either obligatory or optional. The mapping between the surface and deep valences was worked out mostly manually. As a main resource of the Czech surface valences we used the dictionary of Czech surface valences named BRIEF [10] (containing approx. 16,000 verb lemmata). In the BRIEF the verb arguments are typically expressed by noun and prepositional groups in the respective cases using the notation developed for this purpose. Some syntactic transformations can be applied to the valency frames, for instance, passivization. Recently, we have been able to obtain the typical noun and prepositional groups automatically by means of the automatic morphological and syntactic analysis we discussed above.

The described valency frames can serve reasonably well as the descriptions of the meanings of 'legal' verbs – more similar examples can be easily found. For specialized legal verbs, further modifications are needed that require more detailed semantic analysis. To sum up: the goal was to show that valency frames from the VerbaLex database can be appropriately applied to the semantic analysis of the legal language. The size (4830 verbs) of the intersection mentioned above justifies further investigation. We plan to enrich the inventory of the semantic roles in VerbaLex to obtain their more detailed and exact semantic subclassification, or, in other words, more adequate 'legal' ontology.

Then it can be compared with the already existing law ontologies such as the one built within the LOIS (Lexical Ontologies for Legal Information Sharing) project [11][1]. In this project, the ontology was built in the WordNet fashion. However, WordNet-like and similar ontologies are structures capturing relations between nouns and noun groups only. We are convinced that more is needed, in particular, a kind of ontology that can be characterized as 'verbal' [12] in which semantic classes of Czech verbs represent 'verbal' sort of ontology. Semantic roles in the valency frames have served as a criterion for finding relevant semantic classes of (Czech) verbs. This can be also applied to law texts in their natural form. This allows us to conclude that building valency frames of verbs occurring in law texts is one of the important tasks set in the described project.

5   Context Patterns in Law Texts

We believe that the decisive information about the semantics of the law terms can be obtained from the contexts in which they occur. There are two types of approach:

- To use statistical techniques by means of which we obtain the relevant contexts – they can be sorted and the semantic clusters they create can be built. The limitation here is that the data from the law texts are not large enough and in some cases, we do not get enough contexts to make the necessary generalizations.
- To explore the valency frames mentioned above as contexts in the law texts and find the semantic roles in them that are typical for the verbs in the law texts. We have already hinted how this can be done above.

It has to be remarked, however, that even though there are limitations caused by the sparse data, we will investigate intertwining the contexts obtained statistically with the contexts as they are yielded by the valency frames.

6   Conclusion

We have presented the preliminary results of the computational analysis of Czech law documents, or more precisely, their selected samples. On the one hand, we have used the already existing tools such as `ajka` or DIS/VADIS. On the other hand, we have modified them respectively for the purpose of the present task. As a result, we can pay attention to the law language but more importantly, we have obtained a basic knowledge about the grammatical structure of the law texts (and law terminology) and in this way, we are prepared to continue our exploration of the contexts in which law terms occur in the law documents.

In the first part of the paper, we have paid attention to the automatic recognition of the nouns and noun groups occurring in the legal texts – most of them

---

[1] see `http://www.ittig.cnr.it/Ricerca/materiali/lois/WhatIsLOIS.htm` and also `http://nlpweb.kaist.ac.kr/gwc/pdf2006/50.pdf`

serve as law terms. In the second part of the paper, we have concentrated on the verbs and verb groups since they function as relational elements predicting what nouns or noun groups (i. e. terms) will occur together. For capturing and describing argument-predicate structure of verbs, we use the valency frames introduced in the VerbaLex database. The positive and relevant result is that valency frames designed originally for the description of the verbs from general texts are also suitable for capturing the 'legal' verbs. The valency frames can also help us to find contexts that cannot be obtained with statistical techniques because of the low frequency of the 'terminological' verbs in the legal texts.

The knowledge of such contexts is a necessary condition for a deeper understanding of how law terminology works and how it can be made more consistent. As an application, we intend to obtain the basic rules for the intelligent searching of law documents. A tool based on such rules can serve judges, attorneys and experts in creating new law documents. In other words, the relevant output of this work thus will be an electronic dictionary of law terms.

Acknowledgements. This work has been partly supported by the Academy of Sciences of the Czech Republic under the projects 407/07/0679 and by the Ministry of Education of the Czech Republic within the Centre of basic research LC536.

References

1. Sedláček, R.: Morphemic Analyser for Czech. PhD thesis, Faculty of Informatics, Masaryk University, Brno (2005)
2. Šmerk, P.: Towards morphological disambiguation of Czech. PhD thesis proposals, Faculty of Informatics, Masaryk University, Brno (2007) (in Czech)
3. Žáčková, E.: Partial syntactic analysis of Czech. PhD thesis, Faculty of Informatics, Masaryk University, Brno (2002) (in Czech)
4. Horák, A., Hlaváčková, D.: VerbaLex – New Comprehensive Lexicon of Verb Valencies for Czech. In: Computer Treatment of Slavic and East European Languages, Third International Seminar, Bratislava, VEDA, pp. 107–115 (2005)
5. Schulze, B.M., Christ, O.: The CQP User's Manual (1996)
6. Rychlý, P.: Corpus managers and their effective implementation. PhD thesis, Faculty of Informatics, Masaryk University, Brno (2000)
7. Čermák, F., et al.: The Czech National Corpus – SYN2000. Institute of the Czech National Corpus, Prague (2000), http://www.korpus.cz
8. Vossen, P., et al.: The EuroWordNet Base Concepts and Top Ontology. Technical Report Deliverable D017, EuroWordNet LE2 4003, University of Amsterdam (1998)
9. Miller, G.A., Fellbaum, C., et al.: WordNet 3.0. Princeton University (2006), http://wordnet.princeton.edu
10. Pala, K., Ševeček, P.: Valence českých sloves (Valences of Czech Verbs). In: Sborník prací Filozofické fakulty Masarykovy univerzity, Brno, Masaryk University, pp. 41–54 (1997)
11. Peters, W., Sagri, M., Tiscornia, D.: The structuring of legal knowledge in LOIS. Artficial Intelligence and Law 15, 2 (2007)
12. Hlaváčková, D., Khokhlova, M., Pala, K.: Semantic Classes of Czech Verbs. In: Proceedings of the IIS Conference 2009, Krakow (2009) (in print)

Integrating a Bottom–Up and Top–Down Methodology for Building Semantic Resources for the Multilingual Legal Domain

Enrico Francesconi[1], Simonetta Montemagni[2], Wim Peters[3], and Daniela Tiscornia[1]

[1] Institute of Legal Information Theory and Techniques, CNR, Italy
[2] Istituto di Linguistica Computazionale, CNR, Italy
[3] Natural Language Processing Research Group, University of Sheffield, UK

Abstract. This article presents a methodology for multilingual legal knowledge acquisition and modelling. It encompasses two comlementary strategies. On the one hand, there is the top–down definition of the conceptual structure of the legal domain under consideration on the basis of expert jugdment. This structure is language–independent, modeled as an ontology, and can be aligned with other ontologies that capture similar or complementary knowledge, in order to provide a wider conceptual embedding. Another top–down approach is the exploitation of the explicit structure of legal texts, which enables the targeted identification of text spans that play an ontological role and their subsequent inclusion in the knowledge model.

On the other hand, the linguistically motivated, text-based bottom–up population and incremental refinement of this conceptual structure using (semi-)automatic NLP techniques, maximizes the completeness and domain-specificity of the resulting knowledge.

The proposed methodology is concerned with the relation between these two differently derived types of knowledge, and defines a framework for interfacing lexical and ontological knowledge, the result of which offers various perspectives on multilingual legal knowledge.

Two case-studies combining bottom-up and top-down methodologies for knowledge modelling and learning are presented as illustrations of the methodology.

Keywords: Knowledge Modelling, Knowledge Acquisition, Natural Language Processing, Ontology Learning.

1 Introduction

Since the legal domain is strictly dependent on its own textual nature, a methodology for knowledge extraction should take into account a combination of theoretical modelling and text analysis. Such a methodology expresses, in a coherent way, the links between the conceptual characterization, the lexical manifestations of its components and the universes of discourse that are their proper referents.

E. Francesconi et al. (Eds.): Semantic Processing of Legal Texts, LNAI 6036, pp. 95–121, 2010.
© Springer-Verlag Berlin Heidelberg 2010

The aim of this article is, therefore, to set out, through a description of some of the projects that have been implemented, the methodological routes for constructing legal ontologies in applications that, due to the tasks they intend to achieve, should maintain a clear reference to texts. The article is structured in the following way: in Section 2 we analyse the interconnections between language and law and the semantic relations among levels of the legal discourse; in Section 3 we outline the methodological issues inspiring the implementation of the DALOS knowledge modelling and the approach of knowledge acquisition from legal texts; in Section 4 a complementary method for ontology learning is presented dealing with a legal rule learning approach; finally, in Section 5 we comment on the lessons we have learnt.

2   Language and Law

There is a strict connection between law and language, characterised by the coexistence of two autonomous but structurally similar systems: both are endowed with rules that underlie the construction of the system itself, that guide its evolution and guarantee its consistency. Both are conditioned by the social dimension in which are placed, whereby they dynamically define and fix their object in relation to a continually evolving social context.

Law is strictly dependent on its linguistic expression: it has to be communicated, and social and legal rules are mainly transmitted through their written (and oral) expression. Even in customary law there is almost always a phase of verbalisation that enables it to be identified or recognised; even if the law cannot be reduced to language that expresses it, nonetheless, it cannot escape its textual nature.

Another characteristic of law is that it is expresses through many levels of discourse:

- the legislative language is the "object" language because it is the principal source of positive law that, in its broad sense, also includes contracts and so-called soft law; the constitutive force of written sources originates from the stipulative nature of legislative definitions, that assign a conventional meaning of legal concepts in relation to the domain covered by the law that contains them.
- Judges interpret legal language in an 'operative' sense to apply norms to concrete cases: the main function of judicial discourse lies, therefore, in populating the extensional dimension of the object language, instantiating cases throughout judicial subsumption. This involves the linking of general and abstract legislative statements to their linguistic manifestation, or, in others words, the mapping of legal case elements to the kinds of descriptions that may classify them.
- the language of dogmatics is a reformulation of legislative and jurisprudential language aimed at the conceptualisation of the normative contents. Although it is a metalanguage with respect to legislative and judicial language, it is

still based on the analysis of the universe of discourse and it is dependent on specific normative systems.

- At a more abstract level, legal theory expresses the basic concepts, the systemic categories common to (almost) all legal systems (for example, duty, permission, right, liability, sanction, legal act, cause, entitlement etc.). Legal theory may, therefore, be constructed as a formal and axiomatic system, made up of concepts and assertions in the theory, whose truth is based not on a semantic model of reality, but on syntactic rules, derived from inferential, deductive resoning whose scope is explaining positive legal systems [19].
- At the highest level of abstraction, the role of philosophy of law is to express both general principles and value judgements, as well as their ordering criteria.

At the (meta)theoretical level, the border between legal theory and dogmatics may be seen as a genus/species relationship, or as a semantic relation between a logical theory and its models; legal theory has an explanatory and prescriptive function (in the broad sense) because it constructs concepts independently of the normative enunciations and interpretative operations, while the conceptual models of dogmatics arise from the analysis of legal texts, which produces interpreted knowledge.

One of the most obvious areas that demonstrate this distinction is the creativity of legal translation, perched halfway between term equivalence and concept comparison. Legal terminology used in the various legal systems, both European and non-European, expresses not only the legal concepts which operate there, but further reflects the deep differences that exist between the various systems and the different legal perspectives of the lawyers in each system. Given the structural domain specificity of legal language, we cannot speak about "translating the law" to ascertain correspondences between legal terminology in various languages, since the translational correspondence of two terms satisfies neither the semantic correspondence of the concepts they denote nor the requirements of the different legal systems.

Transferred into the computational context, the boundary between the conceptualisations of legal theory and legal concepts built by dogmatics becomes purely methodological. The former entities, the kernel legal concepts, are modelled in the so-called core ontologies, while the latter provide content to domain ontologies, conceived as a possible, non-exclusive interpretation of linguistic objects.

These peculiarities should be taken into account while designing a methodology for ontology construction in the legal domain. They are also relevant for the combination of linguistic and ontological knowledge, in order to best reformulate the process of pure legal scholar conceptualization in a computational context, strongly based on legal language analysis. A general methodology for meaning extraction must be set up within a modular architecture, where different aspects refer to specific analytic models and appropriate Natural Language Processing (NLP) tools.

3   Legal Ontology Construction

Legal ontologies are increasingly becoming a popular field of research, as testified by the list of existing ontologies built for the legal domain which is growing rapidly over the years (for an extensive survey of existing legal ontologies, see [52], [13]). They differ in their purpose or subject-matter, they exhibit varying degrees of generality, formality or richness of internal structure; other relevant differences reflect the methodologies followed for their development, as well as the tools and knowledge representation language used.

Among these different parameters for classifying ontologies, a particularly interesting but often neglected one, deserving, in our opinion, specific attention, concerns the construction process: how was the ontology built? Unfortunately, as pointed out by Paslaru and Tempich [39] in a survey regarding several aspects of ontology development (i.e., methodology and tools used), it appears that "only a small percentage of ontology–related projects follow a systematic approach to ontology building, and even less commit to a specific methodology. Most of the projects are executed in an ad–hoc manner". Most developers do not offer an account of the followed methodological steps, and even when this is the case, it turns out that an ad hoc rather than an established methodology has been followed. This cannot be considered a marginal issue, because it has consequences at different levels on the final result of the construction process.

In this section, we would like to address the methodological issue of how a legal ontology ought to be built. In particular, in Section 3.1 we discuss general methodological issues associated with the construction of an ontology and introduce our approach to legal ontology building. The proposed approach will be illustrated in detail through its implementation within the European DALOS project (in Section 3.2).

3.1   Approaches to Ontology Design and Development

In principle, two different approaches can be recognized as far as the construction of ontologies is concerned: top–down and bottom–up.

In a top–down approach, ontology construction starts by modelling top level concepts, which are then subsequently refined. This approach is typically carried out manually by domain experts and leads to a high–quality engineered ontology. On the other hand, a bottom–up approach to ontology construction starts from the assumption that most concepts and conceptual structures of the domain, as well as the terminology used to express them, are contained in documents. In this approach, the terminological and conceptual knowledge contained in document collections is semi–automatically extracted from texts, thus creating the basis for ontology construction.

There are pros and cons connected with both approaches. Among the advantages usually associated with the top–down construction approach there is the fact that top–down ontologies may be reused across different application scenarios, and can serve as a starting point for developing new ontologies. Among the drawbacks typically associated with top–down ontologies it is worth to mention here that they necessarily require an expert–based approach. Their development is costly in terms of both time and effort. Due to this fact, their coverage is typically rather restricted, and this is a disadvantage when they are used in the framework of real knowledge management applications. Other central problems connected with a top–down approach are the linking of textual information to the ontology, which requires linguistic knowledge about the terminology used to convey domain–specific concepts. Furthermore, there is the highly dynamic and constantly evolving nature of ontologies in different domains, including the legal one, which continuously need to be updated and refined.

When compared with top–down ontology construction, bottom–up approaches have the main advantage of making it possible to discover ontological knowledge at a larger scale and a faster pace; they can also be of some help for detecting and revising human–introduced biases and inconsistencies. Moreover, bottom–up approaches can support the refining and expanding of existing ontologies by incorporating new knowledge emerging from texts. Another crucial aspect is concerned with the fact that they create the prerequisites for the alignment between the ontology and texts: with ontologies boostrapped from texts the linking with textual information is made easier. Among the cons usually ascribed to this class of approaches, there is the fact that a bottom–up approach results in a very high level of detail which makes it difficult to spot commonality between related concepts and increases the risk of inconsistencies [51].

This short characterization of the top–down and bottom–up approaches to ontology construction shows their complementarity. Preferring one approach over the other means ignoring complementary information that can help creating a better product. This fact is more and more acknowledged in the literature, where it is claimed that any comprehensive domain ontology needs work from top–down and bottom–up. Only by proceeding in this way, the resulting ontology reflects domain knowledge and is at the same time anchored to texts. From a general perspective, this is explicitly claimed by Uschold and Grüninger [51], who include among their guidelines for ontology construction and merging the so–called "middle–out approach", based on the combination of top–down and bottom–up ontology modelling. More recently, scholars advocating a middle–out approach to ontology construction started explicitly mentioning the "support of automatic document analysis" through which relevant lexical entries are extracted semi–automatically from available documents (see, for instance, [49]). The (semi–)automatic support in ontology development is nowadays referred to as "ontology learning". Ontology learning represents a promising line of research which is concerned with knowledge acquisition from texts as a basis for the construction and/or extension of ontologies, and in which the learning process is typically carried out by combining NLP technologies with machine learning techniques. Ontology learning is attracting increasing attention as a way to support the task of developing and maintaining ontologies [11] [12].

In the legal domain, the number of ontologies being constructed is rapidly increasing. Most of them still focus on an upper level of concepts and were mostly hand–crafted in a top–down manner by domain experts on the basis of insights from legal theory. More recently, there have been few ontology learning experiments focused on concepts extraction as a primary step of the ontology development process. Among them it is worth mentioning here: the work on definitions in a large collection of German court decisions by [54] [55]; the extraction of domain relevant terminology from normative texts on the basis of which domain relevant concepts are derived together with relations linking them (see, [32], [33], [34]). To our knowledge, relatively few attempts have been made so far to build legal ontologies following a middle–out approach: this is the case, for instance, for the LKIF Core ontology [29], the lexical ontology LOIS [50] [40], the Ontology of Professional Judicial Knowledge [13], and the DALOS ontology [22], where only the latter two appear to resort to ontology learning techniques as far as the bottom–up acquisition process is concerned [41]. Last but not least, a kind of middle–out approach to legal ontology construction is proposed by Saias and Quaresma [46], who exploit NLP tools in order to identify and extract legal concepts and properties: the new domain ontology bootstrapped from texts is then integrated and merged with an externally defined upper foundational legal ontology, with the result of creating a new domain ontology combining low–level concepts with top–level ones.

On the basis of what has been said so far, we believe that the most promising way to build legal ontologies is through the integration of top–down and bottom–up approaches. Such an integrated approach leads to accurate ontology construction, which cannot be achieved by either bottom–up or top–down approach alone. This is particularly true in the legal domain, where ontology construction should follow insights provided by legal theory but at the same time should guarantee textual grounding. Although it is a widely acknowledged fact that ontology building is primarily concerned with the definition of concepts and relations holding between them, it should also include the extraction of linguistic knowledge about the terms used in texts to convey a specific concept, and their relations such as synonymy. In the following section, we will detail how bottom–up and top–down approaches to ontology construction have been combined together into a single construction process in the framework of the DALOS project.

3.2   Knowledge Modelling in the DALOS Project

DALOS[1] was a project launched within the "eParticipation" framework, the EU Commission initiative aimed at promoting the development and use of Information and Communication Technologies in the legislative decision–making processes. The aim of this initiative was to foster the quality of the legislative production, to enhance accessibility and alignment of legislation at European level, and to promote awareness and democratic participation of citizens in the legislative process.

In particular, DALOS aimed to ensure that legal drafters and decision–makers have control over the legal language at national and European level, by providing

---

[1] DrAfting Legislation with Ontology–based Support.

law–makers with linguistic and knowledge management tools to be used in the legislative processes, in particular within the phase of legislative drafting. To this specific end, a knowledge resource was designed and implemented within the project, the DALOS Knowledge Organization System (KOS).

The DALOS KOS is organized in two layers:

- the *Ontological layer*, containing the conceptual modelling at a language–independent level;
- the *Lexical layer*, containing multi–lingual terminology conveying the concepts represented at the Ontological layer.

Concepts at the Ontological layer are linked by taxonomical as well as object property relationships (e.g.has_object_role, has_agent_role, has_value, etc.). On the other hand, the Lexical layer aims at describing the language–dependent lexical expression of the concepts contained in the Ontological layer. At this level, lexical units can be linked through linguistic relationships such as synonymy, hypernymy, hyponymy, meronymy, etc.

In the DALOS KOS, the two layers are connected by relationships mapping concepts to their linguistic counterpart, i.e. terms: this mapping is implemented through the hasLexicalization relationship, which from a monolingual perspective maps a given concept to the term(s) expressing it, whereas from a cross–lingual perspective it maps a given concept to the multilingual terminological variants conveying it.

In this two–layer architecture, the Ontological layer acts as a layer that aligns concepts at the European level, independently from the language and the legal order, where possible. Moreover, the Ontological layer allows to reduce the computational complexity of the problem of multilingual term mapping (N–to–N mapping). Concepts at the Ontological layer act as a "pivot" meta–language in an N–language environment, allowing the reduction of the number of bilingual mapping relationships from a factor N^2 to a factor $2N$. Entries and relationships at both levels are described by exploiting the expressiveness of RDF/OWL semantic Web standards.

The two–level knowledge architecture is illustrated in Figure 1, where it can be noticed that the Ontological layer provides a detailed semantic description of the defined concepts and their relationships and properties, and the Lexical layer describes its linguistic counterpart through the domain terms and the linguistic relationships linking them.

The terms at the Lexical layer are linked by different types of linguistic relationships: for instance, the English term *supplier* is linked to its hyponyms *supplier of goods* and *supplier of services* as well as to its Italian translation equivalent *fornitore*. Another type of lexical relationship, so–called fuzzynym, appears to hold between the terms *consumer* and *supplier*: such a relationship refers to a wider associative relation linking words which may share a number of salient features (in the case at hand, of being involved in a commercial transaction) without being necessarily semantically similar. At the Ontological layer the defined concepts are linked through different types of relationships, namely

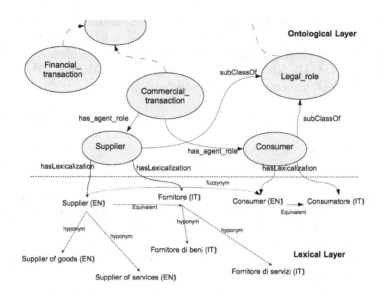

Fig. 1. Knowledge Organization System (KOS) of the DALOS resource

subClassOf (such as the one holding between the SUPPLIER and LEGAL_ROLE concepts) and has_agent_role (linking the COMMERCIAL_TRANSACTION concept to the SUPPLIER one). It is interesting to note that the semantic relatedness between the terms *supplier* and *consumer* captured by the fuzzynym relationship at the lexical level is assigned an explicit semantic interpretation at the ontological level, where it can be noticed that the corresponding concepts a) relate as agents to the COMMERCIAL_ TRANSACTION concept, and b) are subclasses of the LEGAL_ROLE concept.

The DALOS KOS was built following the middle–out approach sketched in Section 3.1. In particular, the DALOS KOS construction was articulated into three main lines of activity:

1. the top–down construction of a (core) domain ontology;
2. the semi–automatic extraction of terminology from domain corpora in different languages by using Natural Language Processing technologies combined with Machine Learning techniques;
3. the refinement of the Ontological and Lexical layers and well as the linking between the two, driven by the terminological and ontological knowledge extracted from the domain corpora.

Whereas the first activity line refers to a top–down process carried out manually by domain experts, the second one corresponds to a bottom–up process aimed at boostrapping the domain terminology from legal document collections. The third activity line refers to the linking of the Ontological layer and the Lexical layer as well as to the refinement of both of them on the basis of the lexical and ontological knowledge bootstrapped from texts. It is at this level that the results of the top–down and bottom–up processes are combined together through an incremental process. For instance, the results of the term extraction process can play an important role by suggesting ontology concepts which were not originally included in the top–down ontology. In principle, the reverse could also hold, in the case where no terms have been acquired that denote some of the concepts included in the Ontological layer.

In what follows, the three activity lines will be illustrated in detail, with particular emphasis on their interaction. Note that for the DALOS case study the "consumer protection" domain has been selected.

Construction of the DALOS Domain Ontology. The Ontological layer of the DALOS resource is aimed at providing an alignment of concepts at language-independent level. It acts not only as a pivot structure for language-dependent lexical manifestations, but it provides an ontologically characterized description of the chosen domain in terms of concepts and their relations, exploiting the expressiveness and reusability of the RDF/OWL semantic Web standards for knowledge representation. This allows also to validate the developed knowledge resource with respect to existing foundational or core ontologies.

As discussed above, the Ontological layer is the result of an intellectual activity aimed at describing the consumer protection domain, chosen for the pilot case. Within the project constraints, an intellectual approach has been chosen to manually capture ontological relations between concepts, relying on expert judgment.

Classes and properties have been implemented on the basis of the terminological knowledge extracted from the chosen Directives on consumer protection law, in particular from the "definitions" contained, maintaining coherence to the design patterns of the Core Legal Ontology (CLO)[2] [25] developed on top of DOLCE foundational ontology [36] and the "Descriptions and Situations" (DnS) ontology [24] [35] within the DOLCE+ library[3]. The DALOS ontology covers the entities pertinent to the chosen domain and their legal specificities. In this knowledge architecture the role of a core legal ontology is to provide entities/concepts, which belong to the general theory of law, bridging the gap between domain-specific concepts and the abstract categories of formal upper level or foundational ontologies such as, in our case, DOLCE.

As regards domain-specific concepts, the DALOS Ontological layer is designed to stress the distinction identified by the "Descriptions and Situations" ontology, extended by CLO within the legal domain, between *intensional specifications* like norms, contracts, roles, and their *extensional realizations* in the same domain, such as cases, contract executions, and agents. This distinction is formally captured by the so called *Norm* ↔ *Case* design pattern [26] (CODeP[4]). According to the *Norm* ↔ *Case* CODeP, *intensional specifications* like norms use tasks, roles, and parameters, while *extensional realizations* like legal cases conform to

---

[2] http://www.loa-cnr.it/ontologies/CLO/CoreLegal.owl
[3] DOLCE+ library, http://dolce.semanticweb.org
[4] Conceptual Ontology Design Pattern.

norms when actions, objects and values are classified by tasks, roles, and parameters respectively. The matching is typically performed when checking if each entity in a legal fact is compliant to a concept in a legal description [26].

The distinction stressed by DALOS is strictly linked to the activity of legislative drafting addressed by the project. Apart from more technical provisions like 'amendments' on existing norms, legislative drafting can in fact be considered as an activity that creates norms on generic situation descriptions, qualifying them by, for example, deontic terms [29]. According to CLO, this activity deals with descriptions (*intensional specifications*) of generic situations (also called "situational frameworks" in [29]), giving them a normative perspective. For example the Directive 97/7/EC of 20 May 1997, at Art. 7 paragraph 1 states that "Unless the parties have agreed otherwise, the supplier must execute the order within a maximum of 30 days from the day following that on which the consumer forwarded his order to the supplier". This states that, unless differently agreed, the generic situation in which the supplier executes an order to the consumer, following a consumer request, is obliged, and this obligation has to be satisfied within a maximum of 30 days from the consumer request.

A normative perspective on generic situations is the result of the legislative drafting activity; it results in legislative text paragraphs grouped into articles, which can be semantically qualified as *provisions* [7], i.e. fragments of a regulation (for example an *obligation* for a role with respect to a task).

A support for legislative drafting can therefore include: 1) a taxonomy of provision types able to give a normative perspective to generic situations; 2) a knowledge resource supporting the description of generic situations in a specific domain, as well as giving an ontological perspective on entities involved in such situations [9]. The DALOS Ontological layer aims at representing this second kind of knowledge resource, tailored to the consumer protection domain pilot case.

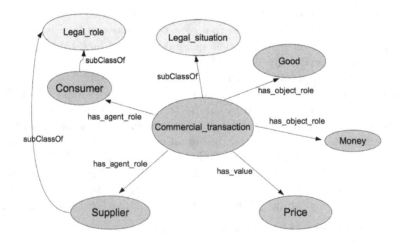

Fig. 2. Excerpt of the DALOS Ontological Layer

The Ontological layer is therefore populated by the conceptual entities, which characterize the consumer protection domain. The first assumption is that all concepts *defined* within consumer law are representative of the domain, and, as a consequence, that several concepts *used* in the definitional contexts pertain to the ontology as well, representing the basic properties or, in other words, the 'intensional meaning' of the relevant concepts. Similarly, the Ontological layer contains generic situations having a legal relevance in the chosen domain.

Such domain-specific concepts are classified according to more general notions, imported from CLO, such as LEGAL_ROLE and LEGAL_SITUATION. Examples of some concepts obtained by the definitions from the consumer law domain are COMMERCIAL_TRANSACTION, CONSUMER, SUPPLIER, GOOD, PRICE. The specific roles they play ([35]) are illustrated in Fig. 2.

On the other hand, the main entities derived from CLO are axiomatized, disjoint classes, characterized by meta properties, such as Identity, Unity and Rigidity. The most relevant distinction is between Roles (anti-rigid) and Types (which are rigid). Roles, according to [35], are anti-rigid since they are "properties that are contingent (non-essential) for all their instances". Types on the other hand can play more roles at the same time. For instance, a legal subject (either a natural or artificial person) can be a seller and a buyer. Domain-specific requirements are expressed by restrictions over ontological classes, for instance by defining CONSUMER as a role that can be *played by* NATURAL_PERSON only.

The first version of the DALOS Ontological layer contains 121 named classes with necessary & sufficient definitions, implemented in the OWL-DL language.

Terminology Extraction in the DALOS Project. Term extraction is the first and most–established step in ontology learning from texts. Terms are surface realisations of domain–specific concepts and represent, for this reason, a basic prerequisite for ontology construction as well as more advanced ontology learning tasks. In principle, they need to be recognized whatever the surface form they show in context, irrespectively of morpho–syntactic and syntactic variants. A term can be a common noun as well as a complex nominal structure with modifiers (typically, adjectival and prepositional modifiers). Term extraction thus requires some level of linguistic pre–processing of texts.

In the DALOS project, term extraction was performed on the English and Italian parts of the DALOS consumer law multi–lingual corpus, including Directives, Regulations and case law on protection of consumers' economic and legal interests. The corpus was built by legal experts and includes 16 Directives and 42 Case Law texts, a total of 292,609 Italian and 273,667 English word tokens.

Term extraction was performed with two different acquisition systems, which were used for dealing with English and Italian texts respectively. For English, GATE[5] [15] was used, a framework for language engineering applications, which supports efficient and robust text processing. GATE uses NLP based techniques to assist the knowledge acquisition process for ontological domain modelling, applying automated linguistic analysis to create ontological knowledge from

---

[5] http://www.gate.ac.uk

textual resources, or to assist ontology engineers and domain experts by means of semi–automatic techniques. For Italian, T2K (Text–to–Knowledge)[16] [34] was used, a hybrid ontology learning system combining linguistic technologies and statistical techniques.

In both cases, term extraction was carried out on the results of a linguistic pre–processing stage, in charge of enriching the original corpus with valuable linguistic information, which is added to the text by means of annotations, in turn used in the subsequent analysis stages. The linguistic pre–processing modules are in charge of:

1. tokenisation of the input text;
2. sentence splitting, segmenting the text into sentential units;
3. morphological analysis (including lemmatisation);
4. part of speech tagging;
5. shallow syntactic parsing (so–called "chunking").

The starting point of the term extraction process is different for the two systems: whereas term extraction in GATE is perfomed against the pos–tagged text (i.e. the output of step 4 above), T2K starts from the shallow parsed text (step 5). To be more concrete, for what concerns English, term candidates are extracted from the text by first selecting either individual pos–tags or sequences of part of speech tags constituting noun phrases, as exemplified below:

– noun (e.g. *creditor, product*);
– adjective–noun (e.g. *current account, local government*);
– noun–noun (e.g. *credit agreement, product safety*);
– noun–preposition–adjective–noun (e.g. *purchase of immovable property, principle of legal certainty*);
– noun–preposition–noun–noun (e.g. *cancellation of credit agreement, settlement of consumer dispute*).

For Italian texts, candidate terms are identified in the shallow parsed texts on the basis of a set of chunk patterns encoding syntactic templates of candidate either simple or complex terms. For what concerns the latter, chunk patterns were defined to cover the main modification types observed in complex nominal terms: i.e. adjectival modification (e.g. *organizzazione internazionale* 'international organisation'), prepositional modification (e.g. *commercializzazione di autovetture* 'marketing of cars'), including more complex cases where different modification types are compounded (e.g. *commercio di prodotti fitosanitari* 'trade of fitosanitary products'). The set of chunk patterns used to identify candidate complex terms was tailored to meet the specific needs of the legal domain, characterised by the frequent use of deep PP-attachment chains including a high number of embedded prepositional chunks [53].

Having identified both single and multi–word term candidates from texts, the following step consists in filtering through the candidates to separate terms from non–terms. This step involves the use of statistically–based measures to compute whether and to what extent a term candidate qualifies as a terminological unit.

In the literature, measures for identifying terms range from raw frequency to Information Retrieval measures such as Term Frequency/Inverse Document Frequency (TF/IDF) [45], the C/NC–value method [23], and lexical association measures such as log likelihood [17], mutual information, or entropy.

In GATE term filtering was perfomed on the basis of the TF/IDF measure, a technique widely used in information retrieval and text mining taking into account term frequency and the number of documents in the collection, and yielding a score that indicates the salience of term candidates for each document in the corpus. All term candidates with a TF/IDF score higher than an empirically determined threshold have been selected: in the DALOS case, a TF/IDF threshold value of 5 yielded 3000 selected terms.

T2K adopts a different term filtering strategy. If on the one hand single terms are identified on the basis of raw frequency in the source document collection (after discounting stop–words), on the other hand multi–word terms are selected on the basis of the log–likelihood measure, an association measure that quantifies how likely the constituents of a complex term are to occur together in a corpus if they were (in)dipendently distributed, where the (in)dependence hypothesis is estimated with the binomial distribution of their joint and disjoint frequencies. The lists of acquired potential single and complex terms are then ranked according to raw frequency and the associated log–likelihood ratio respectively. The selection of the final set of terms (both single and complex ones) requires some threshold tuning, depending on the size of the document collection and the typology and reliability of expected results. In T2K, thresholds define *a)* the minimum frequency for a candidate term to enter the lexicon, and *b)* the overall percentage of terms that are promoted from the ranked lists. For the DALOS corpus, we adopted the following thresholds: minimum frequency threshold equal to 5 for both single and complex terms; selected single terms cover the topmost 20% in the ranked list, whereas selected multi–word terms correspond to the topmost 70% of the ranked list of candidate complex terms. With this configuration, we obtained a term list of 1,443 terms (both single and multi–word terms), of which 1,168 are multi–word terms of different complexity corresponding to the 80% of the acquired term list.[6]

Evaluation of acquired English and Italian term lists was carried out with respect to a subset of 56 of the European Union Legal Concepts (EULG concepts) from LOIS (see [40] and [37] for the complete list) which were selected as a gold standard. The selection of these EULG concepts was based on the fact that they are explicitly listed and defined in the directives included in the DALOS corpus, and are therefore considered to play an important role in their conceptual characterization. Achieved results are promising in both cases: for English, the percentage of correctly acquired terms with respect to all terms appearing in the gold standard terminology is 73.2%, for Italian 80.69%.

---

[6] This peculiar distribution of single vs complex terms follows from the fact that multi–word terms appear to cover the vast majority of domain terminology (85% according to [38]).

For Italian, another evaluation type was carried out, to assess the precision of acquired results, calculated as the percentage of correctly acquired terms with respect to all acquired terms. Automatically acquired terms were evaluated against two reference resources, namely the *Archivio DoGi (Dottrina Giuridica)*[7] and *JurWordNet* [24], containing respectively 9,127 keywords and 5,353 lemmata; note that these resources could not be used for an evaluation in terms of recall (calculated as the percentage of correctly acquired terms with respect to all terms in the reference lexicon) due to their wider coverage, which is not limited to the selected domain. By considering both full and partial[8] matches, the observed precision corresponds to 85.38%, with only 14.62% cases of non–matching terms. Manual inspection of non–matching cases showed that only 6.1% of the cases were to be considered as real errors.

Semi–automatic Refinement and Linking of the Ontological and Lexical Layers. The result of the first two activity lines consists of a hand–crafted core domain ontology and of multilingual term lists. It goes without saying that, when considered separately, the two results cannot effectively be used to support legal knowledge management applications. Only the linking of the domain–specific terms extracted from texts to their description in the ontology provides a usable platform for semantic interpretation of textual information. In this section, we will briefly illustrate the strategy followed within the DALOS project for term to concept mapping, where the results of the bottom–up acquisition process are used both to define the mapping between the Lexical and Ontological layers and to refine the already defined ontology.

First, acquired terms were carefully evaluated by domain experts and linked to the concepts they express in the top–down ontology. It may be the case that newly acquired terms do not find a counterpart at the ontological level; if judged as relevant by domain experts, the ontology is revised accordingly.

However, term extraction is not the only contribution of bottom–up approaches to ontology construction. Extracted terms need to be organized into proto–conceptual relational structures, for them to be exploited in the ontology refinement by domain experts. At this level, different types of relations linking acquired terms can be discovered, based on their distribution in texts.

Starting from the lists of acquired English and Italian terms, different types of lexical relations holding between them were extracted. Acquired relations were in turn used to model and refine the Ontological layer, both at the level of defined concepts and of the relationships linking them.

First, for both English and Italian, the acquired terms were organized into fragments of taxonomical chains, whereby terms such as *time–share contract*,

---

[7] http://nir.ittig.cnr.it/dogiswish/
dogiConsultazioneClassificazioneKWOC.php

[8] Partial matches refer to the following cases: a) the same term appears both in the extracted termbank and in the gold standard resource under different prototypical forms; b) the gold reference resource contains a more general term whereas the extracted list includes one of its hyponyms; c) the gold reference resource contains a more specific term with respect to the extracted list which includes its hypernym.

credit contract and *consumer contract* were classified as co–hyponyms of the general term *contract*. In both cases, taxonomical relationships between terms (typically, single and multi–word terms) were reconstructed by exploiting the internal structure of noun phrases [10]: under this approach, a taxonomic relation is acquired as holding between a single term and all complex terms with this term as the headword.

For English, a second acquisition technique has been experimented with, based on Hearst patterns [28], i.e. a set of lexico–syntactic patterns typically conveying information about hyponymic relations in unrestricted texts. Consider the following example pattern, i.e. "NP such as (NP,)* (or—and) NP" where NP stands for a Noun Phrase and the regular expression symbols have their usual meanings, matching the following context: *advertising and marketing practises, such as product placement, brand differentiation or the offering of incentives . . . .* From contexts like this one it is possible to acquire hyponymic relations such as the one holding between the term *product placement* and the more general term *advertising and marketing practices*. Taxonomical relations acquired with this technique are not limited to head–sharing terms only. Typically, with this technique a high level of precision can be achived, but quite low recall [14]. Unfortunately, this turned out not to be the case with the DALOS corpus; as reported in [41], Hearst patterns appear very rarely in legal corpora.

The identification of taxonomic relations between terms allows the ontology engineer to create concept hierarchies that represent the backbone of the ontology under construction. These linguistic relations can then be reformulated in terms of ontological relations, by means of the OWL SubClassOf relation. Examples from the DALOS ontology are:

DISTANCECONTRACT   SubClassOf CONTRACT
COMMERCIALACTIVITY SubClassOf ACTIVITY

whose linguistic counterpart (namely, *distance contract* is hyponym of *contract* and *commercial activity* is hyponym of *activity*) has been extracted from both the English and Italian corpora.

Yet, taxonomic relations do not exaust the typology of linguistic relations holding between terms which can be automatically extracted from running texts.

For Italian, T2K also acquires clusters of semantically related terms on the basis of distributionally–based similarity measures [1]: following this approach, two terms are semantically related if they can be used interchangeably in a statistically significant number of syntactic contexts. For all terms (both single and complex ones) in the acquired list, we extracted a set of 1,071 semantically related terms referring to 238 terminological headwords. Clusters of automatically acquired semantically related terms are exemplified below:

```
disposizioni 'provision'
   norme, disposizioni legislative, decisione, atto, prescrizioni
legge 'law'
   regolamento, protocollo, accordo, statuto, amministrazioni comunali
```

```
pubblicità ingannevole 'misleading advertisement'
    pratiche commerciali, procedimento, pubblicità comparativa, clausole abusive,
    pubblicità
```

It should be appreciated that in these clusters of semantically related words different classificatory dimensions are inevitably collapsed; they include not only quasi–synonyms (as in the case of *disposizioni* 'provision' and *norme* 'regulations'), hypernyms and hyponyms (e.g. *pubblicità* 'advertisement' and *pubblicità ingannevole* 'misleading advertisement'), but also looser word associations. As an example of the latter we mention the relation holding between *legge* 'law' and *amministrazione comunale* 'municipal administration', or between *comitato* 'committee' and *membri* 'members'.

Acquired clusters of semantically related words can be usefully exploited for the linking between the Lexical and Ontological layers as well as for refining the Lexical and Ontological layers of the DALOS KOS. At the lexical level, whenever possible semantic relatedness between words detected through distributionally–based measures is encoded in terms of lexical paradigmatic relationships such as synonymy, hyponymy/hypernymy, meronymy, antonymy, etc. Remaining relations, which should rather be ascribed to a generic syntagmatic relatedness between words (due to any kind of functional relationship or frequent association), have been encoded in the Lexical layer of DALOS KOS in terms of a rather vague relationship, so–called `fuzzynym` (see Figure 1). For what concerns the Ontological layer, acquired relations have been carefully evaluated by domain experts and encoded in terms of new classes and/or properties (see Figure 1 and its discussion above).

For what concerns English, experiments have been carried out with respect to two syntagmatic relation types, namely a) verbal complementation patterns and b) syntagmatic relations detected through association measures.

Verbal patterns typically reflect lexicalized semantic relations between its arguments. In order to investigate the nature of the semantic contribution from verbal patterns the user needs to be enabled to browse a text according to predefined patterns. Patterns defined in the GATE interface can consist of any type of text annotation that has been added in GATE, e.g. part of speech, string value, lemma etc. The corpus indexing and querying tool in GATE, called ANNIC (ANNotations In Context) [3], allows the evaluator to enter search patterns over text annotations, and detect semantic relations between ontology elements at the fine-grained text level. As proof of concept, the following simple pattern was defined, which identifies pairs of elements from the DALOS ontology that are mentioned in the texts as verb arguments. The surface representation restricts the verb context to a two-token window on either side.

DalosConcept(Token)*2Token.category=="VERB"(Token)*2 DalosConcept

A graphical user interface allows the user to query a corpus and inspect the results from the query. The screenshot in Figure 3 below illustrates how the

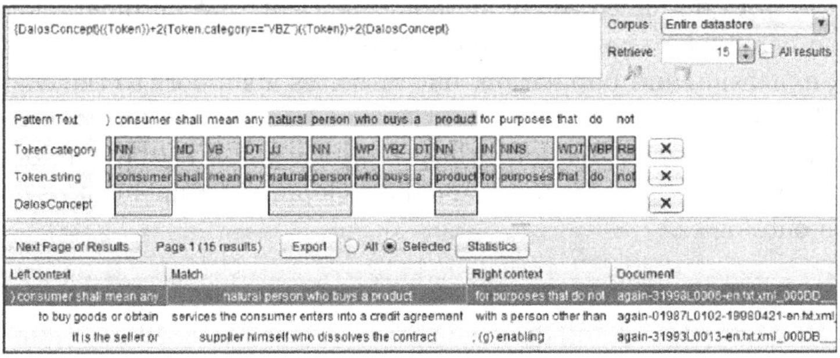

Fig. 3. Snapshot of ANNIC functionality

results are displayed in the GATE interface. Annotations over spans of text are displayed as rows with coloured blocks indicating part of speech, string and DalosConcept. Contexts to the left and right of the text matching the search pattern are displayed at the bottom.

Using this query, 56 patterns were extracted, of which 37 (66%) were evaluated as deserving expert attention. For example:

NaturalPerson conclude Contract   with Seller or Supplier
NaturalPerson buy   Product
Seller/Supplier dissolve Contract
Consumer   enter into CreditAgreement

For what concerns the second experiment, pointwise mutual information (PMI) is a well-known technique that measures the mutual dependence of the two variables as an expression of a syntagmatic relation. It is commonly used as a significance function for the computation of collocations in corpus linguistics [48], measuring the statistically-based strength of relatedness through collocation within the same document. Overall, forty PMI relations were found between existing concepts from the DALOS ontology after matching DALOS ontology labels onto textual elements. Nine (22.5%) of the forty are not connected by any relation or concatenation of relations in the ontology. Consider, for instance, the following pairs with their associated PMI value:

CONSUMERGOODS   CONSUMERPROTECTION 4.10099
CONSUMERPROTECTION CONSUMER   3.37321
FINANCIALSERVICE   ˙ SUPPLIER   2.56943

It turned out that 77.5% of the extracted MI relations are already attested in the ontology. The 22.5% of the MI pairs without ontological confirmation make ontological sense in that they express fine–grained relations that should be expertly evaluated for inclusion into the ontology, and linked to existing ontology elements by means of existing or new object properties. In general, the significant overlap between pointwise mutual information results and existing ontological relations indicate the relevance of such a measure for ontology acquisition.

From the work discussed so far, it should be clear that automatic knowledge acquisition cannot be seen as a stand–alone method for ontology creation, refinement, expansion and population, but rather as a support to the engineering activity of domain experts. In this section, we have shown how the results of text–driven knowledge extraction, which is just a phase in the ontology development cycle, can be used for the manual development, refinement or extension of domain ontologies.

4   Legal Rules Learning

In this section an approach to knowledge acquisition in the legal domain is presented, which is complementary to the DALOS methodology.

Domain ontologies assume a specific importance in the legal domain since they provide knowledge, in terms of concepts and their relationships, on scenarios addressed by legal rules, expressed in legal texts. Domain concepts addressed by legal rules are particularly relevant for the legal domain. In fact, in this domain users are mainly interested in accessing concepts regulated by norms. They look for legal reasoning and consultancy support, as, for example, instruments to check compliance with procedures with respect to specific statutes and regulations.

The approach presented in this section addresses the identification of domain concepts addressed by legal rules, as derived from knowledge extraction techniques, aimed at legal rules learning from legislative texts. The extracted domain concepts as well as the established relationships can represent a starting point for the implementation of domain ontologies.

An approach to support the acquisition of legal rules contained in legislative documents has been recently proposed [8] [21]. It is based on a semantic model for legislation and implemented by using knowledge extraction techniques over legislative texts. This methodology is targeted at providing a contribution to bridge the gap between consensus and authoritativeness in legal rule representation, because it contributes to reaching consensus by limiting human intervention in the descrituion of legal rules, which are extracted from authoritative texts as the legislative ones.

The proposed approach to legal knowledge acquisition is based on learning techniques targeted at extracting legal rules from text corpora. Legal rules are essentially "speech acts" [47] expressed in legislative texts regulating entities of a domain: their nature therefore justifies an approach aimed at the analysis of such texts. Therefore, the proposed knowledge acquisition framework is based on a twofold approach:

1. Knowledge modelling: definition of a semantic model for legislative texts able to describe legal rules;
2. Knowledge acquisition: instantiation of legal rules through the analysis of legislative texts, being driven by the defined semantic model.

This approach traces a framework which combines top-down and bottom-up strategies: a top–down strategy provides a model for legal rules, while a bottom-up strategy identifies rules instances from legal texts. The bottom–up knowledge acquisition strategy in particular can be carried out manually or automatically. The manual bottom-up strategy consists, basically, of an analytic effort in which all the possible semantic distinctions among the textual components of a legislative text are identified. On the other hand, the automatic (or semi-automatic) bottom-up strategy performs the previous activities with support from automatic tools that are able to classify rules, according to the defined model, and identify the involved entities.

The knowledge model proposed in this work reflects this orientation and is organized into the following two components:

1. Domain Independent Legal Knowledge (DILK)
2. Domain Knowledge (DK)

DILK is a semantic model of Rules expressed in legislative texts, while DK is any terminological or conceptual knowledge base (thesaurus, ontology, semantic network) able to provide information and relationships among the Entities of a regulated domain. The combination of DILK with one or more DKs is able to provide a formal characterization of Rules instances. For this reason the proposed methodology to legal knowledge modelling has been called *DILK-DK* approach [21].

DILK. DILK is conceived as a model for legal Rules, independently from the domain they apply to. In literature several models (classification) of legal rules have been proposed, from the traditional Hohfeldian theory of legal concepts [30] until more recent legal philosophy theories due to Rawls [43], Hart [27], Ross [44], Bentham [6], Kelsen [31].

In this respect, the work of Biagioli [7] deserves particular attention. Combining the work of legal philosophers on rules classification with the Searlian theory of rules preceived as "speech acts", as well as the Raz's lesson [42] to perceive laws and regulations as a set of *provisions* carried by speech acts, Biagioli underlined two views or *profiles* according to which a legislative text can be perceived: a) a structural or *formal profile*, representing the traditional legislator habit of organizing legal texts into chapters, articles, paragraphs, etc.; b) a semantic or *functional profile*, considering legislative texts as composed by *provisions*, namely fragments of regulations [7] expressed by speech acts. Therefore, a specific classification of legislative provisions was carried out by analysing legislative texts from a semantic point of view, and grouping provisions into two main families: *Rules* (introducing and defining entities or expressing deontic concepts) and *Rules on Rules* (different kinds of amendments). Rules are provisions which aim at regulating the reality considered by the including act. Adopting a typical law theory distinction, well expressed by Rawls, rules consist of:

- *constitutive rules*: they introduce or assign a juridical profiles to entities of a regulated reality;
- *regulative rules*: they discipline actions ("rules on actions") or the substantial and procedural defaults ("remedies").

On the other hand, Rules on Rules can be distinguished into:

- *content amendments*: they modify the literal content of a norm, or their meaning without literal changes;
- *temporal amendments*: they modify the times of a norm (come-into-force and efficacy time);
- *extension amendments*: they extend or reduce the cases on which the norm operates.

In Biagioli's model each provision type has specific arguments describing the roles of the entities which a provision type applies to (for example the *Bearer* is argument of a *Duty* provision). *Provision types* and related *Arguments* represent a semantic model for legislative texts [7]. They can be considered as a sort of metadata scheme able to analytically describe fragments of legislative texts. For example, the following fragment of the Italian privacy law:

> *"A controller intending to process personal data falling within the scope of application of this act shall have to notify the "Garante" thereof, ... "*

besides being considered as a part of the physical structure of a legislative text (a *paragraph*), can also be viewed as a component of the logical structure of it (a *provision*) and qualified as a *provision* of type *Duty*, whose arguments are:

Bearer: "Controller";	*Object*:	"Process personal data"
Action: "Notification"	*Counterpart*: "Garante"	

The specific textual anchoring of Biagioli's model represents, in our opinion, its main strength. Since the DILK-DK approach aims at representing Rules instances as expressed in legislative texts, we consider Biagioli's model, limited to the group of rules, as a possible implementation of DILK. "Rules on rules" affect indirectly the way how the reality is regulated, since they amend Rules in different respects (literally, temporarily, extensionally): therefore such provision types are not part of DILK model. On the other hand, their effects on Rules has to be taken into account for knowledge acquisition purposes.

DK. In legislative texts *Entities* regulated by provisions are expressed by lexical units. These can be provided by a *Domain Knowledge* (DK) repository providing conceptualization of entities consisting of the language-dependent lexical units[9]. Information on such entities at language-independent level, as well as their lexical manifestations in different languages needs to be described by DK. A possible architecture for describing DK has been proposed within the DALOS project[10]

---

[9] "Typically regulations are not given in an empty environment; instead they make use of terminology and concepts which are relevant to the organisation and/or the aspect they seek to regulate. Thus, to be able to capture the meaning of regulations, one needs to encode not only the regulations themselves, but also the underlying ontological knowledge. This knowledge usually includes the terminology used, its basic structure, and integrity constraints that need to be satisfied." [2].

[10] http://www.dalosproject.eu

(see Section 3.2). More details on the DALOS DK architecture, as well as a possible implementation of it for the domain of consumer protection, can also be found in [22] (see also previous section).

Knowledge Acquisition. Knowledge acquisition within the DILK-DK framework consists of two main steps: 1) DILK instantiation, 2) DK construction.

DILK instantiation. The DILK instantiation phase is a bottom-up strategy for legislative text paragraphs classification into *provision types*, as well as specific lexical units identification, assigning them roles in terms of *provision arguments*. The automatic bottom-up strategy, here proposed, consists in using tools able to support the human activity of classifying provisions, as well as to extract their arguments. Three main steps can be foreseen:

− Collection of legislative texts and conversion into an XML format [5]
− Automatic classification of legislative text paragraphs into provisions [20]
− Automatic argument extraction [8]

Legislative documents are firstly collected and transformed into a jurisdiction-dependent XML standard (NormeInRete in Italy, Metalex in the Netherlands, etc.). For the Italian legislation a module called xmLegesMarker, of the xmLeges[11] software family, has been developed [5]. It is able to transform legacy content into XML in order to identify the formal structure of a legislative document.

For the automatic classification of legislative text paragraphs as provison types, a tool called xmLegesClassifier of the xmLeges family has been developed. xmLegesClassifier has been implemented using a Multiclass Support Vector Machine (MSVM) approach, which provided the best results in preliminary experiments compared to other machine learning approaches [20]. With respect to [20], in this work MSVM is tested on the Rules provision family, as the first step of DILK instantiation[21].

A tool called xmLegesExtractor[12] [8] of the xmLeges family has been implemented for the automatic detection of provision arguments. xmLegesExtractor is realized as a suite of NLP tools for the automatic analysis of Italian texts (see [4]), specialized to cope with the specific stylistic conventions of the legal parlance. A first prototype takes as input legislative raw text paragraphs, coupled with the categorization provided by the xmLegesClassifier, and identifies text fragments (lexical units) corresponding to specific semantic roles, relevant for the different types of provisions (Fig. 4). The approach follows a two–stage strategy. The first stage consists in a syntactic pre–processing which takes in input a text paragraph, which is tokenized and normalized for dates, abbreviations and multi–word expressions; the normalized text is then morphologically analyzed and lemmatized, using an Italian lexicon specialized for the analysis of legal

---

[11] http://www.xmleges.org
[12] xmLegesExtractor has been developed in collaboration with the Institute of Computational Linguistics (ILC-CNR) in Pisa (Italy).

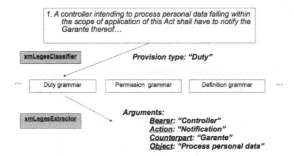

Fig. 4. xmLegesClassifier combined with the grammar approach used by xmLegesExtractor

language; finally, the text is POS-tagged and shallow parsed into non–recursive constituents called "chunks" [18]. The second stage consists in the identification of all the lexical units acting as arguments relevant to a specific provision type. It takes in input a chunked representation of legal text paragraphs, locating relevant patterns of chunks which represent entities with specific semantic roles within a provision type instance, by using a specific provision type oriented grammar (Fig. 4).

DK construction. Lexical units identified by xmLegesExtractor represent language-dependent lexicalizations of provision arguments. More information on related entities, as well as their relations within a specific domain, can be obtained by mapping lexical units to concepts in any existing DK repositories. On the other hand, the extracted information can be considered as a basis on which to construct DK repositories (in terms of thesauri or domain ontologies). Actually, their construction is not a specific task of *legal* ontologists, but of ontologists *tout court*, since a DK repository has to contain information on entities of a domain independently from a legal perspective. This aspect is important in order to conceive a legal knowledge architecture whose components can be reused. A DILK-DK learning approach only suggests that only language-dependent lexical units are contained in DK repositories, which can be implemented by projecting lexical units onto a large text corpora of a specific domain, inferring conceptualizations by term clustering, as well as using statistics on recurrent patterns for discovering term relationships. This issue is out of the paper scope; a vast literature exists on this topic, therefore the interested reader can refer to [12].

Benefits of the DILK-DK bottom-up learning approach. The proposed learning approach for legal knowledge acquisition can provide the following benefits: a) it assists the implementation of taxonomies, or suggests concepts for hand-crafted ontologies [55]; b) it contributes to bridging the gap between authoritativeness and consensus for legal rule representation, since it is able to extract rules directly form legislative texts, which are authoritative sources (by definition), while promoting consensus, since rules are automatically extracted from legal sources, limiting human interaction.

5   Discussion

In analysing legal documents, the first aspect that must be elicited is the relation between meaning (norm) and form (text). Norms are conceived as the interpreted meaning of written regulations that correspond to a partition in legal text, such as articles and paragraphs. Additionally, a norm can be built by interpretative activities on a set of linguistic expressions logically entailed, for instance, the decision in a judgement, or set of legislative statements in a judgement, or set of legislative statements. Only in few cases (definitions, deeming provisions) legal concepts are elicited from the core meaning of a single norm, but more frequently they are built on sets of norms, through a process of abstraction and generalization, by collecting sets of normative conditions, to be linked to sets of legal effects.

To respect the peculiarities of the legal domain, different approaches should be adopted in the process of legal concept extraction. On one side, a conceptual model of the domain needs to be created, either by means of manual ontology engineering or the extraction of the intensional definition of legal concepts from linguistic contexts. Legislative definitions are generally expressed by fixed linguistic structures within a legislative text, and therefore they can be easily identified and isolated.

On the other side, the analysis of text containing legislative provision instances can identify relevant concepts as well as relationships pertaining to a regulated domain, thus providing effective hints for the construction of a domain ontology as well as linking the related concepts to core and fundamental ontologies.

Techniques such as term extraction, lexical analysis, parsing and statistical collocations (as discussed and illustrated in previous sections) yield textually derived information, which can then be re-engineered into ontological concepts, concept properties and relations. The work performed in both the DALOS project and the approach followed within the DILK-DK framework is illustrative of this type of activity, by means of bottom–up knowledge acquisition in the former case, or as a result of provisions categorization in the latter case.

The application of both top–down and bottom–up knowledge acquisition techniques to the legal domain enables the adoption of several perspectives on legal knowledge and the formulation of various aims within the legal field. For example, in the field of legal comparison, different conceptualizations (resulting from the bottom-up analysis of texts from different legal systems) can be compared throughout a shared reference ontology, conceived as a level of abstraction, which legal experts can agree upon. In addition, the same model can be exploited in European law-making, where conceptual equivalence of multiligual entities is assumed. The role of a reference ontology is, in this case, to assess (multilingual) terminological consistency, i.e. whether different lexicalization reflects the same normative conceptual meaning.

References

1. Allegrini, P., Montemagni, S., Pirrelli, V.: Example-Based Automatic Induction of Semantic Classes Through Entropic Scores. In: Linguistica Computazionale, XVI-XVII, Tomo I, pp. 1–45 (2003)
2. Antoniou, G., Billington, D., Governatori, G., Maher, M.: On the modeling and analysis of regulations. In: Proceedings of ACIS, pp. 20–29 (1999)
3. Aswani, N., Tablan, V., Bontcheva, K., Cunningham, H.: Indexing and Querying Linguistic Metadata and Document Content. In: Proceedings of 5th International Conference on Recent Advances in Natural Language Processing, Borovets, Bulgaria (2005)
4. Bartolini, R., Lenci, A., Montemagni, S., Pirrelli, V., Soria, C.: Automatic classification and analysis of provisions in italian legal texts: a case study. In: Meersman, R., Tari, Z., Corsaro, A. (eds.) OTM-WS 2004. LNCS, vol. 3292, pp. 593–604. Springer, Heidelberg (2004)
5. Bacci, L., Spinosa, P., Marchetti, C., Battistoni, R.: Automatic mark–up of legislative documents and its application to parallel text generation. In: Proceedings of LOAIT Workshop, Barcelona, Spain, pp. 45–54 (2009)
6. Bentham, J., Hart, H.L.A.: Of Laws in General (1st edn., 1872). Athlone, London (1970)
7. Biagioli, C.: Towards a legal rules functional micro–ontology. In: Proceedings of LEGONT (1997)
8. Biagioli, C., Francesconi, E., Passerini, A., Montemagni, S., Soria, C.: Automatic semantics extraction in law documents. In: Proceedings of ICAIL, pp. 133–139 (2005)
9. Biagioli, C., Cappelli, A., Francesconi, E., Turchi, F.: Law making environment: perspectives. In: Proceedings of the V Legislative XML Workshop, pp. 267–281. European Press Academic Publishing, Firence (2007)
10. Buitelaar, P., Olejnik, D., Sintek, M.: A protégé plug–in for ontology extraction from text based on linguistic analysis. In: Bussler, C.J., Davies, J., Fensel, D., Studer, R. (eds.) ESWS 2004. LNCS, vol. 3053, pp. 31–44. Springer, Heidelberg (2004)
11. Buitelaar, P., Cimiano, P., Magnini, B.: Ontology Learning from Text: an Overview. In: Buitelaar, et al. (eds.) Ontology Learning from Text: Methods, Evaluation and Applications. Frontiers in Artificial Intelligence and Applications, vol. 123, pp. 3–12 (2005)
12. Buitelaar, P., Cimiano, P. (eds.): Ontology Learning and Population: Bridging the Gap between Text and Knowledge. IOS Press, Amsterdam (2008)
13. Casellas, N.: Modelling Legal Knowledge through Ontologies. OPJK: the Ontology of Professional Judicial Knowledge. PhD thesis, Faculty of Law, Universitat Autònoma de Barcelona (2008)
14. Cimiano, P., Pivk, A., Schmidt-Thieme, L., Staab, S.: Learning taxonomic relations from heterogeneous sources. In: Proceedings of the ECAI 2004 Ontology Learning and Population Workshop (2004)
15. Cunningham, H., Maynard, D., Bontcheva, K., Tablan, V.: Gate: A framework and graphical development environment for robust nlp tools and applications. In: Proceedings of the 40th Anniversary Meeting of the Association for Computational Linguistics, ACL 2002 (2002)

16. Dell'Orletta, F., Lenci, A., Marchi, S., Montemagni, S., Pirrelli, V., Venturi, G.: Dal testo alla conoscenza e ritorno: estrazione terminologica e annotazione semantica di basi documentali di dominio. In: Atti del Convegno Nazionale Ass.I.Term I-TerAnDo, Università della Calabria, 5-7 giugno 2008, Roma, AIDA Informazioni, n. 1-2/2008, pp. 185–206 (2008),
    `http://www.aidainformazioni.itpubdellorletta-atal122008.pdf`
17. Dunning, T.: Accurate methods for the statistics of surprise and coincidence. Computational Linguistics 19(1), 61–74 (1993)
18. Federici, S., Montemagni, S., Pirrelli, V.: Shallow Parsing and Text Chunking: a View on Underspecification in Syntax. In: Proceedings of the Workshop on Robust Parsing, European Summer School on Language, Logic and Information (ESSLLI 1996), Prague (1996)
19. Ferrajoli, L.: Principia iuris Laterza Bari (2007)
20. Francesconi, E., Passerini, A.: Automatic classification of provisions in legislative texts. International Journal on Artificial Intelligence and Law 15, 1–17 (2007)
21. Francesconi, E.: An Approach to Legal Rules Modelling and Automatic Learning. In: Proceedings of the JURIX Conference (2009)
22. Agnoloni, T., Bacci, L., Francesconi, E., Peters, W., Montemagni, S., Venturi, G.: A two-level knowledge approach to support multilingual legislative drafting. In: Breuker, J., Casanovas, P., Klein, M.C.A., Francesconi, E. (eds.) Law, Ontologies and the Semantic Web. Channelling the Legal Information Flood. Frontiers in Artificial Intelligence and Applications, vol. 188, pp. 177–198. IOS Press, Amsterdam (2009)
23. Frantzi, K., Ananiadou, S.: The C-value/NC-value domain independent method for multiword term extraction. Journal of Natural Language Processing 6(3), 145–179 (1999)
24. Gangemi, A., Sagri, M.T., Tiscornia, D.: Jur-wordnet, a source of metadata for content description in legal information. In: Proceedings of the ICAIL Workshop on Legal Ontologies & Web based legal information management (2003)
25. Gangemi, A., Sagri, M.T., Tiscornia, D.: A constructive framework for legal ontologies. In: Benjamins, V.R., Casanovas, P., Breuker, J., Gangemi, A. (eds.) Law and the Semantic Web. LNCS (LNAI), vol. 3369, pp. 97–124. Springer, Heidelberg (2005)
26. Gangemi, A.: Design patterns for legal ontology construction. In: Casanovas, P., Noriega, P., Bourcier, D. (eds.) Trends in Legal Knowledge. The Semantic Web and the Regulation of Electronic Social Systems, pp. 171–191. European Press Academic Publishing, Firence (2007)
27. Hart, H.: The Concept of Law. Clarendon Law Series. Oxford University Press, Oxford (1961)
28. Hearst, M.A.: Automatic acquisition of hyponyms from large text corpora. In: Proceedings of the 14th International Conference on Computational Linguistics, pp. 539–545 (1992)
29. Hoekstra, R., Breuker, J., Bello, M.D., Boer, A.: The lkif core ontology of basic legal concepts. In: Casanovas, P., Biasiotti, M., Francesconi, E., Sagri, M.T. (eds.) Proceedings of the Workshop on Legal Ontologies and Artificial Intelligence Techniques. CEUR Workshop Proceedings, vol. 321, pp. 43–63 (2007) ISSN 1613-0073,
    `http://CEUR-WS.org/`
30. Hohfeld, W.N.: Some fundamental legal conceptions. Greenwood Press, Westport (1978)
31. Kelsen, H.: General Theory of Norms. Clarendon Press, Oxford (1991)

32. Lame, G.: Knowledge acquisition from texts towards an ontology of French law. In: Proceedings of the International Conference on Knowledge Engineering and Knowledge Management Managing Knowledge in a World of Networks (EKAW 2000), Juan-les-Pins (2000)

33. Lame, G.: Using NLP techniques to identify legal ontology components: concepts and relations. In: Benjamins, V.R., Casanovas, P., Breuker, J., Gangemi, A. (eds.) Law and the Semantic Web. LNCS (LNAI), vol. 3369, pp. 169–184. Springer, Heidelberg (2005)

34. Lenci, A., Montemagni, S., Pirrelli, V., Venturi, G.: Ontology learning from Italian legal texts. In: Breuker, J., Casanovas, P., Klein, M.C.A., Francesconi, E. (eds.) Law, Ontologies and the Semantic Web, Frontiers in Artificial Intelligence and Applications, vol. 188, pp. 75–94. IOS Press, Amsterdam (2009)

35. Masolo, C., Vieu, L., Bottazzi, E., Catenacci, C., Ferrario, R., Gangemi, A., Guarino, N.: Social roles and their descriptions. In: Welty, C. (ed.) Proceedings of the Ninth International Conference on the Principles of Knowledge Representation and Reasoning, Whistler (2004)

36. Masolo, C., Gangemi, A., Guarino, N., Oltramari, A., Schneider, L.: Wonderweb deliverable d18: The wonderweb library of foundational ontologies. Technical Report (2004)

37. Montemagni, S., Marchi, S., Venturi, G., Bartolini, R., Bertagna, F., Ruffolo, P., Peters, W., Tiscornia, D.: Report on Ontology learning tool and testing. In: Progetto Europeo DALOS (Drafting Legislation with Ontology–Based Support). Deliverable 3.3. DALOS Project (2007)

38. Nakagawa, H., Mori, T.: Automatic Term Recognition based on Statistics of Compound Nouns and their Components. Terminology 9(2), 201–219 (2003)

39. Paslaru, E., Tempich, C.: Ontology Engineering: A Reality Check. In: Meersman, R., Tari, Z. (eds.) OTM 2006. LNCS, vol. 4275, pp. 836–854. Springer, Heidelberg (2006)

40. Peters, W., Sagri, M.T., Tiscornia, D.: The Structuring of Legal Knowledge in LOIS. In: Proceedings of 10 th International Conference of Artificial Intelligence and Law (ICAIL 2005), Bologna, June 6-11 (2005)

41. Peters, W.: Text–based Legal Ontology Enrichment. In: Proceedings of LOAIT 2009, 3rd Workshop on Legal Ontologies and Artificial Intelligence Techniques joint with 2nd Workshop on Semantic Processing of Legal Texts, Barcelona, Spain, June 8, pp. 55–66 (2009)

42. Raz, J.: The Concept of a Legal System, 2nd edn. Clarendon Press, Oxford (1980)

43. Rawls, J.: Two concepts of rule. Philosophical Review 64, 3–31 (1955)

44. Ross, A.: Directives and Norms. Routledge, London (1968)

45. Salton, G., Buckley, C.: Term–Weighting Approaches in Automatic Text Retrieval. Information Processing and Management 24(5), 513–523 (1988)

46. Saias, J., Quaresma, P.: A Methodology to Create Legal Ontologies in a Logic Programming Based Web Information Retrieval System. In: Benjamins, V.R., Casanovas, P., Breuker, J., Gangemi, A. (eds.) Law and the Semantic Web. LNCS (LNAI), vol. 3369, pp. 185–200. Springer, Heidelberg (2005)

47. Searle, J.: Speech Acts: An Essay in the Philosophy of Language. CUP, Cambridge (1969)

48. Smadja, F.A., McKeown, K.R.: Automatically extracting and representing collocations for language generation. In: Proceedings of ACL 1990, Pittsburgh, Pennsylvania, pp. 252–259 (1990)

49. Sure, Y., Studer, R.: A methodology for ontology-based knowledge management. In: Davies, J., Fensel, D., van Harmelen, F. (eds.) Towards the Semantic Web. Ontology-driven Knowledge Management, pp. 33–46. John Wiley & Sons, LTD, Chichester (2003)

50. Tiscornia, D.: The LOIS project: Lexical ontologies for legal information sharing. In: Biagioli, C., Francesconi, E., Sartor, G. (eds.) Proceedings of the V Legislative XML Workshop, pp. 189–204. European Press Academic Publishing, Firence (2007)

51. Uschold, M., Grüninger, M.: Ontologies: Principles, methods, and applications. Knowledge Engineering Review 11(2), 93–155 (1996)

52. Valente, A.: Types and Roles of Legal Ontologies. In: Benjamins, V.R., Casanovas, P., Breuker, J., Gangemi, A. (eds.) Law and the Semantic Web. LNCS (LNAI), vol. 3369, pp. 65–76. Springer, Heidelberg (2005)

53. Venturi, G.: Legal Language and Legal Knowledge Management Applications. In: Francesconi, E., Montemagni, S., Peters, W., Tiscornia, D. (eds.) Semantic Processing of Legal Texts. LNCS (LNAI), vol. 6036, pp. 3–26. Springer, Heidelberg (2010)

54. Walter, S., Pinkal, M.: Automatic extraction of definitions from german court decisions. In: Proceedings of the International Conference on Computational Linguistics (COLING 2006), Workshop on Information Extraction Beyond The Document, Sidney, pp. 20–28 (2006)

55. Walter, S., Pinkal, M.: Definitions in court decisions – automatic extraction and ontology acquisition. In: Breuker, J., Casanovas, P., Klein, M.C.A., Francesconi, E. (eds.) Law, Ontologies and the Semantic Web. Channelling the Legal Information Flood. Frontiers in Artificial Intelligence and Applications, vol. 188, pp. 95–113. IOS Press, Amsterdam (2009)

Ontology Based Law Discovery

Alessio Bosca and Luca Dini

CELI s.r.l., 10131 Torino, Italy
{alessio.bosca,dini}@celi.it

Abstract. The vast amount of information freely available on the Web constitutes a unparalleled resource for automatic knowledge discovery and learning. In this article we propose a study on Ontology Induction for individual laws based on corpora comparison that exploits a domain corpus automatically generated from the Web; in particular we present a case study on the Italian "Legge Bassanini" (59/1997, 127/1997 - concerning the simplification and decentralization of administrative procedures).

We evaluate how the induced ontological characterizations might vary according to different factors, such as the genre (e.g. news vs. social media),the learning algorithm, the text analysis granularity, etc; the main contribution of the paper consists of highlighting the structural difference emerging from the learned predicates, and in showing how the learning mechanism might provide valuable information on how laws are perceived in different layers of the civil society.

Keywords: Ontology Learning, Natural Language Processing, Terminology Extraction.

1 Introduction

The vast amount of information that nowadays is freely available on the Web facilitates the harvesting of information about any domain. The availability of such a resource stimulated an intense research effort in the Ontology Learning field, investigating approaches and algorithms capable of automatically extracting knowledge from the Web and building domain ontologies.

The major challenges in mining the Web for knowledge arise from its inherent unstructured nature: a typical web document presents different, even unrelated chunks of information on the same page (i.e. misleading metadata, information related to site navigation or banner and advertisings). A further element to be taken into consideration, when analyzing web corpora, consists of the differences among the different information sources with respect to style, flavours and reliability of the retrieved data . Two corpora concerning the same domain topic, but based on distinct information sources, such as newspapers v.s. forums, are bound to induce different ontological characterizations, more denotative and descriptive in the first case, more connotative, also reflecting users opinions, in the second case.

E. Francesconi et al. (Eds.): Semantic Processing of Legal Texts, LNAI 6036, pp. 122–135, 2010.
© Springer-Verlag Berlin Heidelberg 2010

In this paper[1] we present a study on Ontology Induction for individual laws based on corpora comparison and exploiting a domain corpus automatically generated from the Web.

We make the case of the Italian law "Bassanini" (59/1997, 127/1997 - concerning the simplification and decentralization of administrative procedures, see [16]) and, by analyzing an automatically generated collection of web documents, we evaluate how the associated induced ontological characterizations might vary according to different factors, such as the genre (e.g. news vs. social media), the learning algorithm, the text analysis granularity, etc.

For our purposes, by "ontology" we do not mean a formalized legal ontology in the sense of, e.g. Peters & al. 2007 or Shaheed, & al. 2005 (see [17] and [21]), but rather a linguistically induced concept cloud which encompasses a set of both logical relations (hypernymy, part-of, use-for, etc.) and purely linguistic relations (prototypical argument, semi fixed expression, etc.). What the system concretely learns is not a model, but a set of predicateswhich are associated with the individual (the Bassanini Law) and for which no explicit formal representation is available.

The main contribution of the article consists of highlighting the structural difference emerging from the learned predicates, and in showing how the learning mechanism might provide valuable information on how laws are perceived at different layers of the civil society.

The paper is organized as follows: Section 2 presents relevant related work; Section 3 presents the approach used for automatically generating the corpus exploited for the statistical and lexical-semantic analyses. Section 4 describes the strategy adopted for ontology induction and Section 5 presents the obtained results, along with some considerations on law perception that emerges from the data. Eventually Section 6 concludes the paper and proposes some future works.

2   Related Work

The research challenge faced by Ontology Learning consists in automatically extracting information from texts and giving a structured organization to the discovered knowledge. Such a topic lies in the convergence of the research activities of Artificial Intelligence, Natural Language Processing and Machine Learning communities; therefore, although a relatively new research topic, ontology learning builds on top of a large amount of previous work and has generated a considerable research effort in the last years (see [1] and [11]).

Paul Buitelaar & al. (in [1]) divided and structured the Ontology Learning activities in the so-called Ontology Learning layer cake, where each layer represents a subtask with an increasing level of knowledge abstraction and respectively involving the identification and organization of terms, synonyms, concepts, concepts hierarchies, concepts relations and rules. The work presented in this paper

---

[1] This work is partially funded by ICT4LAW, a research project financed by "Regione Piemonte".

focuses on the first three layers of the cake and consists in automatically extracting a set of domain terms from a corpus of web documents and interrelating them with linguistic and logical relations.

Our approach to terminology extraction is based on frequencies comparison between a domain corpus and reference corpora; a well known idea in the field with different works addressing the issue (see [4]). Our solution exploits the Log Odds Ratio measure (LOR) in order to evaluate the specificity of a term for the domain under analysis (see [2]) and is similar to the approach proposed in [3].

In conjunction with LOR, we also experimented with a corpus-based distributional approach capable of detecting the interrelation between the extracted terms; the strategy we adopted is similar to Latent Semantic Analysis (see [5]) although it uses a less expensive computational solution based on the Random Projection algorithm (see [6] and [7]). Different works debates on similar issues: [8] uses LSA in order to solve synonymy detection questions from the well known TOEFL test while the methods presented by [10] or by [9] propose the use of the Web as a corpus to compute mutual information scores between candidate terms.

The study presented in this paper investigates the differences emerging from the outcomes of the Ontology Learning process when using different domain corpus as input or different selections of a specific corpus. D. Manzano-Macho & al. in [12] presented a work on the use and integration of different informative sources in an Ontology Learning environment although they focus on the pourpose of enhancing the reliability of information extracted through multisource evidence.

3   Corpus Generation

The corpus we used for the study presented in this paper has been automatically generated from the Web with the aim of collecting documents with a coarse–grained classification about the genre of the contained information (blogs, news, forums) and the date of publication. In order to retrieve such additional information along with the document content we exploited the Google search services on blogs, news and groups (see [14], [15] and [13]).

This solution allows us to discriminate the retrieved documents on the basis of the genre; furthermore the document snippets provided within the search result pages of the Google search services contain additional information associated with each document, such as the date of publication and the name of the blog, group or newspaper.

Our approach to automatic corpus building comprises three different phases: as a first step the domain documents are harvested from the web along with the contextual information provided by the search services; then the documents are post-processed by filtering away off-topic and duplicated data; finally the contents are enriched with linguistic features (lemmatization, named entities recognition), thus creating the final domain corpus.

In the first phase, a simple query containing a reference to the individual law(s) under analysis (consisting of its name and its identificative number) is submitted to the search services; in our case study the query consisted of: "Bassanini 59/1997 127/1997" and returned a total of about 3000 documents. The search result pages are then analyzed in order to extract the contextual data along with the document contents. The common presentation framework adopted by Google groups and blogs for structuring web pages contents allows for an easy detection of metadata such as user names and citations. By contrast, newspaper documents retrieved by the search service originate from heterogeneous sources and do not share any presentation strategy; as a consequence the news documents thus collected contain a certain amount of non–relevant data that should be removed.

In the second phase the retrieved documents are analyzed in order to filter away noisy information and post duplicates. In the context of our case study the noisy information is constituted by off-topic data such as blog/forum user signatures and mottos or navigation information (e.g "click here", "subscribe to this forum", etc.); the duplicates, instead, mainly fall into two different categories: repeated citations that occur within the same document (a very common practice in discussion threads) and different snapshot of the same blog at different dates but nonetheless containing the same post that typically appears in the documents with different order positions.

In the third phase, a shallow linguistic analysis is performed on the retrieved documents in order to lemmatize the textual contents (by associating with each term its lemma and its part of speech) and extract the named entities (such as persons, places and organisations).

At the end of this process the generated corpus comprised a total of 1567 documents and in particular 699 from groups, 181 from blogs and 687 from news.

4 Ontology Learning Approach

The Ontology Learning solution proposed in this paper exploits the corpus automatically harvested from the Web (see Section 3) in order to identify a set of terms that are highly distinctive of the investigated domain; the extracted terms are then interrelated by means of a statistical analysis of words occurrences within the corpus. The resulting outcome consists of a structured representation (although not a formal ontology) of the key concepts within the domain.

Our strategy for Term Extraction is based on corpus frequency comparison between the domain corpus and a general one (reference or background corpus); as term specificity measure our approach exploits a modified version of the well known Log Odds Ratio (LOR, see [2]). The term specificity measure function adopted in our experiments is, in fact, a weighted combination of the LOR and the plain Term Frequency measure. It can be formalized as:

$$TermSpec = k * \frac{TermDF*GC_{Docs}}{TermGF*DC_{Docs}} + TermDF * (1 - k)$$

where TermDF represents the frequency of a given term in the domain corpus, TermGF its frequency in the general corpus, DC_Docs the number of documents comprised in the domain corpus while GC_Docs the number of documents contained in the reference corpus. We experimented the terminology extraction with 3 different value of k (0, 0.5 and 1) resulting therefore in 3 different measure functions: a pure LOR measure (whit k=1) a pure TF measure (with k=0) and an evenly weighted LOR/TF (with k=0.5); the following section describes in details the different outcomes resulted from the adoption of these different measure functions.

Then the terms thus extracted from the domain corpus are enriched with semantically related terminology by means of a corpus-based distributional model; such technology is based on the assumption that the meaning of a given term implicitly emerges from the different contexts where it appears (here with context we intend unit of text as a paragraph, a document or a textual window).

The Random Indexing (RI) exploits an algebraic model in order to represent the semantics of terms in a Nth dimensional space (a vector of length N). The RI approach in fact creates a Terms By Contexts matrix where each row represents the degree of memberships of a given term of the different contexts. The RI algorithm assigns a random signature to each context (a highly sparse vector of length N, with few, randomly chosen, non zero elements) and then generates the vector space model by performing a statistical analysis of the documents in the domain corpus and by accumulating on terms rows all the signatures of the contexts where terms appear.

According to this approach if two different terms have a similar meaning they should appear in similar contexts (within the same documents or surrounded by the same words), resulting charachterized by close coordinates within the thus generated semantic space. In our case study we applied the RI technique for generating term clusters by selecting in the semantic space the terms with the minimal distance from the word under analysis exploiting the well known cosine distance measure.

5   Discovery

5.1   Corpus Footprint

The first step of our discovery experiment is to understand whether we can produce a general "snapshot" of the contents of the corpus and, as a side effect, of the law and its impact on open medias. In order to perform such a task, the results of LOR applied to the document base appear somehow deceptive. Indeed both terms list based on pure LOR (Table 1) or the weighted LOR/TF list (weight 0.5, Table 2) seem to put emphasis on extremely unexpected terms rather than relevant descriptive terms .

We notice of course the appearence of some terms, which are likely to characterize the Bassanini law or opinions people can have about it (e.g. *federalismo, decentramento, amministrazione, federalista, vitalizio*), but the trend is obfuscated by terms which are just unexpected and are probably brought in by some off-topic discussion (*pastrano, montá, tombino*). In order to minimize the impact of off-topic and noise coming from poor structural analysis of web pages, we

Table 1. Pure LOR

Term	Eng. translation	Value
ayatollá	ayatollá	0,99549305
accessiblitá	accessibility	0,97528327
pastrano	greatcoat	0,94161826
incazzerá	get pissed (future)	0,88756603
magiá	magic	0,85293996
rimbambimento	becoming dumb	0,82466036
consacrerá	consecrate	0,8225496
montá	*misspelled word*	0,81579334
responsabilitá	responsibility	0,7787663
parchimetro	parking meter	0,7617317
lavativo	idler	0,7588993
leggittimitá	legitimacy	0,7501239
vitalizio	annuity	0,743974
penzá	*misspelled word*	0,72753
radiotelefono	radio-telephone	0,7266631
neo	mole	0,72025836
dilapidazione	squandering	0,7084538
chioccia	sitting-hen	0,70516056
cartucciera	noun	0,69469285
cotillon	cotillon	0,6864572
tabacchino	tobac shop	0,68518096
leguleio	lawyer	0,68395275
produzione	production	0,6719629

restrict the LOR algorithm *to only sentences containing the word bassanini*. The weighted focused LOR algorithm produce the results reported in Table 3.

The list starts to shed some light on the kind of data we are dealing with, and at least we notice the emergence of a set of terms concerning the direct impact on the citizen (*fotocopia, certificato, documento, delega, firma*) and a set of terms more related to the political impact of the law (*federalismo, decentramento, governo, riforma, ministero*, etc.). We take this set of words as an input seed to our study and we try to see what are the connotation these terms acquire in the corpus under study.

5.2   Comparison among LOR and RI

Once we have isolated a set of concepts, which constitute the pivot of our study, we can investigate different behaviours of the two algorithms. At the same time we can evaluate the effect of linguistic phenomena such as semantic ambiguity and syntactic constituency on the proposed methodology.

The analysis of word clustering based on the above seeds (Table 3) seems to confirm that the law is mentioned/discussed either with respect to the very specific aspects which impact citizen life (i.e. interaction with bureaucracy) or on national/political effect such as its connection to *federalism* (which is by the way

Table 2. Weighted LOR

Term	Eng. translation	Value
neo	mole	0,7534129
pastrano	greatcoat	0,7385227
ignobiltá	ignobility	0,7156902
federalismo	federalism	0,7042161
scuola	school	0,6693464
punto	point	0,66918105
legge	law	0,661939
decentramento	decentralization	0,6618327
eriditá	legacy	0,65985215
produzione	production	0,6591317
montá	*misspelled word*	0,6332109
tombino	manhole	0,60322976
consacrerá	consacrate	0,59377927
vitalizio	annuity	0,5917658
fintanto	up to when	0,5833467
amministrazione	administration	0,5824593
centrosinistra	moderate left	0,5820025
federalista	federalist	0,5805112
radioamatore	noun	0,5766138
leghista	supporter of an Italian federalist party called "Lega"	0,5714757

only a marginal part of the law). To see this, we can take the cloud of concepts that RI associates to the word *documento* (see Table 4).

Here it is clear that the concept of "document" is interpreted not in law oriented terms, but it is almost always seen in terms of relations with the actions a citizen has to perform with respect to the public administration. It is quite interesting to compare this cluster with the analogous one obtained with LOR (see Table 5).

We notice that in the LOR analysis the characterization of the law as a tool for citizens in their relationship with the public administration is even stronger, with the enphasis on very specific concepts such as *anagrafe, residenza, cittadinanza* etc. In a sense, this helps to identify the real impact in citizens life. On the other hand we notice that results from LOR excludes all terms that are related to the automation of public administration, which are, on the contrary, present in RI (*informatico, digitale, chiave*): this is a typical effect of the "background" corpus (cf. Section 4) which has been chosen to run LOR: being a web corpus ([3]), terms from the IT field are quite frequent, and as a consequence their high occurrence in the bassanini corpus reduces their importance.

When analysing more institutional/political aspects, the divergence between the two algorithms tends to diminish, probably because of the fact that these topics are less popular in forums driven by non professionals. So, for instance, the most prominent terms associated to *decentramento* by both algorithms are *federalismo, amministratore, secessione*, etc.

Table 3. Weighted Focused LOR

Term	Eng. translation	Value
federalismo	federalism	0,7090571
decentramento	decentralization	0,65736
ipocrita	hypocrite	0,65083504
amministrazione	administration	0,60097665
soppressione	suppression	0,59534335
senato	senate	0,5837337
firma	sign	0,5818429
governo	government	0,5722294
fotocopia	photocopy	0,55560046
seno	breast	0,54869735
prefetto	prefect	0,54598284
semplificazione	simplification	0,54140204
documento	document	0,5251423
sparirá	disappear	0,5239263
asserzione	assertion	0,5235616
delega	proxy	0,5185608
riforma	reform	0,5097327
consulenza	advice	0,5005262
ministero	ministry	0,49301475
intenzione	intention	0,4929664

Table 4. RI associations to the word *documento*

Term	Eng. translation	Value
firma	sign	0,8304054
informatico	IT technician	0,74902385
validitá	validity	0,7096739
scrittura	writing	0,6449456
requisito	requisite	0,6277224
direttiva	directive	0,61613995
efficacia	effectiveness	0,61371666
sottoscrittore	subscriber	0,57873505
sottoscrizione	subscription	0,5768877
legislatore	legislator	0,5651688
riproduzione	reproduction	0,54222584
conformitá	conformity	0,5385759
disciplina	discipline	0,52908784
equiparazione	legal equivalence	0,5144153
prova	proof	0,49302307
digitale	digital	0,4639772
chiave	key	0,46162057

Table 5. LOR associations to the word *documento*

Term	Eng. translation	Value
notorietá	fame	0,5946464
fotocopia	photocopy	0,5866991
firma	sign	0,58326226
spedizione	dispatch	0,5270494
anagrafe	registry office	0,52157855
legge	law	0,49294257
clinica	clinic	0,4788969
identitá	identity	0,47815886
certificato	certificate	0,45043084
residenza	residency	0,43732113
dichiarazione	declaration	0,43013296
cittadinanza	citizenship	0,4107147
atto	act	0,40642926
nascita	birth	0,3901822

One significant aspect that emerges from both LOR and RI, is the occurrence of the nouns *prefetto* (prefect) and *soppressione* (abolition) (not shown in the table): this is a typical case where text discovery methods can help to detect a secondary aspect of the law (actually of a part of the law which was suppressed in the definitive version) which had a strong influence on opinion leaders, be they journalist or bloggers. This fact also sheds light on the need for syntactic analysis to make sense out of these approaches. Indeed, the high occurrence in concept clusters of the words *prefetto* and *soppressione* is related with the fact that the system at this stage lacks the ability to recognize *soppressione del prefetto* as a single chunk. If we were able to detect this syntagmatic chunk we would enrich our conceptual representation, as the semantic role of *patient* would be made syntactically explicit, rather than left to the analyst's intuition and verification. Other cases are even more evident when we pass from the discovery of noun-noun correspondences to the ones syntactically associating items such as verb-nouns or nouns-adjectives. This effect will be examined in depth in Section 5.4.

We observe that, in the clusters of words under analysis, semantic ambiguity does not represent a major problem as the clusterization process tends to isolate the meaning that is most prominent in the corpus. The only exception we observed is related to the word *delega* which relates both to the flow of law approval and to one aspect of the delegation process ruled by the law Bassanini.

5.3   Understanding Groups Attitudes

The main goal of this paper is to understand how different aspects of a law are perceived by the public opinion. However, "public opinion" is not a monolithic block, and we might be interested in understanding which concepts are more prominent for different populations. In our experiment we exploit the distinction between press (controlled information) and consumer produced media (blogs, forums newsgroup: spontaneous information). We apply the same algorithms described above and we compare the associated word clusters.

The first fact which emerges is that certain target words tend to disappear from certain partitions of the corpus, i.e. there are concepts which are only relevant to press and concepts which are found only on blogs. The second one is that for comparatively populated clusters the overlap among clusters is minimal (about 36% on average at the first 20 terms associated with each target word). This is expected, as journals tend to describe the law, while citizens are mostly (but not exclusively) concerned with the usage of the law in practical contexts. However, as expected as it might be, this divergence provides a clear measure of the impact of different parts of a law on different actors of the civil society.

One interesting case of this divergence is represented by the word *certificazione* (certification): in the press section it originates from a quite precise cloud of related concepts, which mirrors, in a sense, the structure of the law, whereas on in open sources section it is associated just with the concepts of *modulo*(form) and *ricorso* (recourse) (Table 6).

In the case of *certificazione* we are facing a kind of *objective data*, in the sense that it is unlikely that an opinion or a subjective attitude on *certificazione* can emerge in either of the two sources. However, if we shift our attention to

Table 6. Cluster for *certificazione*

Term	Eng. translation	Value
Cluster in open newspapers		
autenticazione	authentication	0,60632765
sottoscrizione	subscription	0,5804886
documento	document	0,57164943
certificato	certificate	0,5346612
legislatore	legislator	0,5279385
dichiarazione	declaration	0,5155602
firma	signature	0,5124954
validitá	validity	0,46791664
chiave	key	0,46115094
documentazione	documentation	0,45426464
requisito	requisite	0,45349178
obbligo	duty	0,44990215
atto	act	0,44650328
procedura	procedure	0,44436824
posta	post	0,44169757
sottoscrittore	subscribe	0,43611416
direttiva	directive	0,42388856
utilizzo	use	0,4200472
notorietá	fame	0,41560116
Cluster in open source media		
ricorso	appeal	0,6485791
modulo	application form	0,5787199

Table 7. Cluster for *cittadino*

Term	Eng. translation	Value
Cluster in newspapers		
amministrazione	administration	0,75139165
rapporto	relation	0,6752472
obiettivo	objective	0,62253135
servizio	service	0,6107529
qualitá	quality	0,5980857
diritto	law	0,5917262
possibilitá	possibility	0,58259714
processo	trial	0,5790534
iniziativa	initiative	0,5768706
procedura	procedure	0,5764878
atto	act	0,5719806
Cluster in open source media		
pratica	practice	0,52855957
ente	agency	0,4888266
stato	state	0,47593874
servizio	service	0,47420365
senso	meaning	0,42983234
amministrazione	administration	0,42628178
obbligo	obligation	0,41650316
certificato	certificate	0,40642056
problema	problem	0,404583
tempo	time	0,39950722

concepts which might give rise to subjective attitudes, we observe a clear objec-
tive(press)/subjective(blogs) contrast. Let's take for instance the word *cittadino*.
The associated clouds in press and blogs are presented in Table 7. Here again
press articles provide a quite objective portrait of the intent of the law with
respect to the citizen. The cloud implicitly asserts that the administration provides services to the citizens, it aims at increasing quality, enhancing rights and
opening possibilities. Blogs and newsgoups, on the contrary, put the emphasis
on the practical aspects of the interaction between the state or organization
(not the administration) and the citizen (the noun *pratica* (dossier) here is to
be interpreted as a "beaurocratic step"). Difficult aspects of such an interaction
are emphasized by the following nouns with negative connotations: *problema*
(problem) and *obbligo* (obligation), whereas *pratica*, *certificato* (certificate) and
tempo (time) tend to put emphasis on procedural aspects of the relationship.

5.4   Ontological Relations

The methodology we put in place can be seen as a way of exploring a corpus, characterizing citizen attitudes, but also identifying concepts which might
be related in a manually coded ontology. For instance, if we analyze the cloud

Table 8. Ontological Relations

Pivot Term	Eng Pivot	Related Term	Eng Related	Value
certificato	certificate	richiedere	require	0,302
certificato	certificate	tradurre	translate	0,289
certificato	certificate	utilizzare	use	0,255
certificato	certificate	rimanere	remain	0,254
certificato	certificate	considerare	regard	0,252
certificato	certificate	ritenere	deem	0,244
certificato	certificate	interessare	interest	0,241
certificato	certificate	firmare	sign	0,240
processo	process	consentire	allow	0,424
processo	process	avviare	start	0,395
procedimento	process	simplify	verb	0,511
procedimento	process	streamline	verb	0,282
amministrazione	administration	proporre	propose	0,508
amministrazione	administration	fare	make	0,496
amministrazione	administration	presentare	present/offer	0,489
amministrazione	administration	portare	bring	0,483
amministrazione	administration	require	verb	0,466
firma	sign	autenticare	authenticate	0,348
firma	sign	consentire	allow	0,309
firma	sign	richiedere	require	0,306
firma	sign	recare	bring	0,305
documento	document	allegare	enclose	0,491
documento	document	evitare	avoid	0,328
documento	document	sottoscrivere	subscribe	0,320
documento	document	firmare	sign	0,316
documento	document	richiedere	require	0,313
documento	document	apporre	append	0,306
delega	delegate	abrogare	repeal	0,463
delega	delegate	prevedere	provide for	0,449
delega	delegate	introdurre	introduce	0,401
delega	delegate	modificare	modify	0,385
delega	delegate	approvare	approve	0,381
delega	delegate	emanare	emanate	0,380

above connected to *certificazione* (table 8) we see that different kinds of onto-logical relations are included. For instance a certification **is_a** kind of *document*, **has_part** *signature*, **has_feature** *validity* etc. Of course the characterization of these relations in semantic terms is left to lexicographers, but the proposed algorithm can help in identifying the range of the semantic attributes. We also notice that the proposed algorithm can provide useful hints not only for coding ontologies with respect to nominal concepts, but also for building frame-based ontologies such as a juridical framenet (see [19] or [20]). In order to obtain such results, we restrict neighbour selection to a specific category and we apply the RI algorithm with a limited syntactic window. Table 8 shows the result when applying this method on noun-verbs pairs: it is clear that in most cases the discovery algorithm identifies the verbal contexts where a certain *artifact* acts as a prototypical objects/subjects. For instance, a *document* is typically *required* by an administration, *attached* to a dossier, *signed* and *subscribed* by someone etc. We estimate that this notion of "prototypicality" might reveal extremely useful information for modelling abstract knowledge about administrative procedures, as well as fixing selectional restrictions in domain specific semantic networks.

6   Conclusions and Future Work

In this article we described a methodology, which allows a fast exploration of a domain corpus (in this case cointaining documents on the Bassanini law) while emphasizing at the same time different attitudes of different opinion groups towards the object under study (the law). As a side effect of this exploration, we are able to produce fragments of a corpus oriented informal ontology, which might represent a useful bootstrap for a legal ontologist, or be used "as it is" in some applications such as document classification, information retrieval, etc. The future steps of our research will consist of improving the methodology in such a way that syntactically well formed phrases (and not only single words) can count as targets of the analysis. Moreover, we will explore methods to link these informal ontologies to more formalized ones, such as LOIS (see [17] or [18]) or Jur-WordNet (see [19]).

References

1. Buitelaar, P., Cimiano, P., Magnini, B.: Ontology learning from text: methods, evaluation and applications. IOS Press, Amsterdam (2005)
2. Everitt, B.: The Analysis of Contingency Tables, 2nd edn. Chapman and Hall, Boca Raton (1992)
3. Baroni, M., Bernardini, S.: BootCaT: Bootstrapping corpora and terms from the web. In: Proceedings of LREC 2004 (2004)
4. Rayson, P., Garside, R.: Comparing corpora using frequency profiling. In: Proceedings of Workshop on Comparing Corpora of ACL 2000, pp. 1–6 (2000)
5. Deerwester, S., Dumais, S.T., Furnas, G.W., Thomas, K.L., Harshman, R.: Indexing by latent semantic analysis. Journal of the American Society for Information Science (1990)
6. Bingham, E., Mannila, H.: Random projection in dimensionality reduction: applications to image and text data. In: Proceedings of the Seventh ACM SIGKDD International Conference on Knowledge Discovery and Data Mining (2001)
7. Lin, J., Gunopulos, D.: Dimensionality Reduction by Random Projection and Latent Semantic Indexing. In: Proceedings of the Text Mining Workshop, at the 3rd SIAM International Conference on Data Mining (2003)
8. Turney, P.D.: Mining the Web for Synonyms: PMI-IR versus LSA on TOEFL. In: Proceedings of the Twelfth European Conference on Machine Learning, pp. 491–502 (2001)
9. Baroni, M., Bisi, S.: Using cooccurrence statistics and the web to discover synonyms in a technical language. In: Proceedings of LREC 2004 (2004)

10. Inkpen, D.: Near-Synonym Choice in an Intelligent Thesaurus. In: Proceedings of NAACL/HLT 2007 (2007)
11. Maedche, E., Staab, S.: Ontology Learning for the Semantic Web. IEEE Intelligent Systems (2001)
12. Manzano-Macho, D., Gomez-Perez, A., Borrajo, D.: Unsupervised and Domain Independent Ontology Learning. Combining Heterogeneous Sources of Evidence. In: Proceedings of LREC 2008 (2008)
13. http://groups.google.it
14. http://blogsearch.google.it/
15. http://news.google.it/archivesearch
16. Bassanini, F.: Overview of Administrative Reform and Implementation in Italy: Organization, Personnel, Procedures and Delivery of Public Services. International Journal of Public Administration, 229–252 (2000)
17. Peters, W., Sagri, M.T., Tiscornia, D.: The structuring of legal knowledge in LOIS. Artificial Intelligence and Law 15(2), 117–135 (2007)
18. Mommers, L., Dini, L., Liebwald, D., Peters, W., Schweighofer, E., Voermans, W.J.M.: Cross–lingual legal information retrieval using a WordNet architecture. In: Proceedings of the International Conference on Artificial Intelligence and Law. ACM, New York (2005)
19. Sagri, M.T., Tiscornia, D., Bertagna, F.: Jur-WordNet. In: GWC 2004 Proceedings, pp. 305–310 (2004)
20. Boella, G., van der Torre, L.: A Foundational Ontology of Organizations and Roles. In: Baldoni, M., Endriss, U. (eds.) DALT 2006. LNCS (LNAI), vol. 4327, pp. 78–88. Springer, Heidelberg (2006)
21. Shaheed, J., Yip, A., Cunningham, J.: A top-level language-biased legal ontology. In: Proceedings of the ICAIL Workshop on Legal Ontologies and Artificial Intelligence Techniques, LOAIT (2005)

Multilevel Legal Ontologies

Gianmaria Ajani[1], Guido Boella[2], Leonardo Lesmo[2], Marco Martin[2],
Alessandro Mazzei[2], Daniele P. Radicioni[2], and Piercarlo Rossi[3]

[1] Dipartimento di Scienze Giuridiche - Università di Torino
[2] Dipartimento di Informatica - Università di Torino
[3] Dipartimento di Studi per l'Impresa e il Territorio,
Università del Piemonte Orientale
`gianmaria.ajani@unito.it`,
{`guido,lesmo,mazzei,radicion`}`@di.unito.it`,
`notmart@gmail.com`,
`piercarlo.rossi@eco.unipmn.it`

Abstract. In order to manage the conceptual representation of European law we have proposed the Legal Taxonomy Syllabus (LTS) and
the related methodology. In this paper we consider further issues that
emerged during the testing and use of the LTS, and how we took them
into account in the new release of the system. In particular, we address
the problem of representing interpretation of terms besides the definitions occurring in the directives, the problem of normative change, and
the process of planning legal reforms of European law. We show how
to include into the Legal Taxonomy Syllabus the Acquis Principles -
which have been sketched by scholars in European Private Law from the
so-called Acquis communautaire -, how to take the temporal dimension
into account in ontologies, and how to apply natural language processing
techniques to the legal texts being annotated in the LTS.

Keywords: Multilingual Legal Ontologies, Formal Ontologies, European
Directives.

1   Introduction

European Union Directives (EUDs) are sets of norms that have to be implemented by the national legislations and translated into the language of each
Member State. The general problem of multilingualism in European legislation
has recently been addressed by using linguistic and ontological methodologies
and tools, e.g. [1,2,3,4,5]. The management of EUD is particularly complex,
since the *implementation* of a EUD does not correspond to a straight transposition into a national law. Conversely, managing this kind of complexity with
appropriate tools can facilitate the comparison and harmonization of national
legislation [6]. For instance, the LOIS Project aimed at extending EuroWordnet with legal information, thus adopting a similar approach to multilingualism,
with the goal of connecting a legal ontology to a higher level ontology [4].

E. Francesconi et al. (Eds.): Semantic Processing of Legal Texts, LNAI 6036, pp. 136–154, 2010.
© Springer-Verlag Berlin Heidelberg 2010

In previous work we proposed the Legal Taxonomy Syllabus[1] (LTS), a methodology and a tool to build multilingual conceptual dictionaries aimed at representing and analysing terminologies and concepts from EUDs [8,9,10,11]. The LTS is basically concerned with assisting legal experts in the access to EU documents. LTS is based on the distinction between *terms* and *concepts*. The latter ones are arranged into ontologies that are organised in levels. In [11] only two levels were defined: the European level –containing only one ontology deriving from EUDs annotations–, and the national level –hosting the distinct ontologies deriving from the legislations of EU member states.

While annotating the EUDs, testing and using the system, some requirements emerged from users expert in law, demanding for a more sophisticated approach along with further developmental efforts: first, we noted that it is frequent the case that concepts are the result of a doctrinal interpretation process rather than of a definition in directives. If, on the one hand, the definitions in directives and their relation with the actual text are required by legal scholars to have a precise model of European law, the layman is more interested in the concepts which result from the doctrinal interpretation. Second, laws are typical objects evolving through time. Another open issue to cope with in building legal frameworks both at the European and at the national level is the *normative change* [12,13]. Concepts in the legal ontologies should not only represent the consolidated legal text, but should also keep trace of the evolution of the meaning. Finally, besides the actual directives, the European Union aims to harmonize law by reformulating terminology in a more coherent way. The European Commission provide in various ways common principles, terminology, and rules for law to address gaps, conflicts, and ambiguities emerging from the application of European law, and this effort should be taken into account in the LTS as well.

Thus, in this article we address the following research questions:

- How to consider not only the terms defined in the directives but also the interpretation process of legal scholars in the LTS? How to better integrate concepts and the text of EUDs in the LTS?
- How to extend the ontology with a temporal dimension to be able to represent normative change? How to allow users to search also for past meanings of terms and the modified norms introducing them?
- How to extend the levels of the LTS from European and national to new ones representing the possible reforms of European law? How to represent the relation between the existing European law and the planned revisions?

We answer the first question by introducing concepts called *abstract*, in that they are not related to a single directive, which should be conveniently recognized as a *grouping* of concepts. The users will be allowed to navigate the ontology at different levels of detail, depending on their goals. Moreover, we use

---

[1] LTS is a dictionary of Consumer Law, which has been carried out within the broader scope of the Uniform Terminology Project, http://www.uniformterminology. unito.it [7]. The implemented system can be found at the URL: http://www. eulawtaxonomy.org

natural language processing techniques to facilitate the management of legal text associated with concepts.

We answer the second question by introducing *time* into the ontology and by allowing to have new concepts replacing the old ones while keeping the latter in the system as well.

We answer the third question by introducing new levels in the LTS and show how can they manage a set of principles (namely, the *Acquis Principles*), which is gaining in popularity, with the aim of improving the quality of EUDs national implementations by Member States.

Note that our answers concern the methodology underlying LTS, but all the features that will be described have a direct software implementation in the LTS tool. The paper is structured as follows. In Section 2 we introduce the new legal requirements we take into account in the revision of the LTS. In Section 3 we first recall the LTS (Section 3.1) and then we explain how the new system satisfies the additional requirements previously outlined (Sections 3.2 to 3.5). The article ends with conclusions.

2   Multilingual and Multilevel Ontologies for European Directives

In this Section we start by briefly summarizing the motivations which lead to the development of the LTS (Section 2.1). Then we introduce new requirements which have been raised by legal experts using the LTS (Sections 2.2 to 2.4).

2.1   Terminological and Conceptual Misalignment

Comparative Law has identified two key points in dealing with EUDs, which address the polysemy of legal terms: we call them the *terminological* and *conceptual misalignments*.

In the case of a EUD (usually adopted for harmonising the laws in the Member States), the terminological matter is complicated by the need to implement it in the national legislations. In order to have a precise transposition into a national law, a Directive may be subject to further interpretation. A single concept in a particular language can be expressed in a number of different ways in a EUD and in the national law implementing it. As a consequence we have a terminological misalignment. For example, the concept corresponding to the word *reasonably* in English, is translated into Italian as *ragionevolmente* in the EUD, and as *con ordinaria diligenza* in the transposition law.

In the EUD transposition laws a further problem arises from the different national *legal doctrines*. A legal concept expressed in a EUD may not be present in a national legal system. In this case we can talk about a conceptual misalignment. To make sense for the national lawyers' expectancies, the European legal terms need not only to be translated into a sound national terminology, but also to be correctly detected when their meanings refer to EU legal concepts or when their meanings are similar to concepts which are known in the Member states. Consequently, the transposition of European law in the parochial legal framework of each Member state can lead to a set of distinct national legal doctrines, that are all different from the European one. In the case of consumer contracts (like those concluded by means of distance communication, as in Directive 97/7/EC, Art. 4.2), a related example of this phenomenon concerns the notion of the professionals providing in a *clear and comprehensible manner* some elements of the contract to the consumers, which represents a specification of the information duties that are a pivotal principle of EU law. Despite the pairs of translation in the language versions of EU Directives (e.g., *klar und verständlich* in German - *clear and comprehensible* in English - *chiaro e comprensibile* in Italian), each legal term, when transposed in the national legal orders, is influenced by the conceptual filters of the lawyers' domestic legal thinking. So, *klar und verständlich* in the German system is considered by the German commentators as referring to three different legal concepts: 1) the print or the writing of the information must be clear and legible (*Gestaltung der Information*), 2) the information must be intelligible by the consumer (*Formulierung der Information*), 3) the language of the information must be the national language of the consumer (*Sprache der Information*). In Italy, the judiciary tend to control the formal features of the criterions 1 and 3, and apply criterion 2 to a lesser extent, while in England the emphasis is on criterion 2, although this is interpreted as plain style of language (and not as legal technical jargon) because of the historical influences of the plain English movement in that country.

Note that this kind of problem identified in comparative law has a direct correspondence in the ontology theory. In particular Klein [14] has identified two particular forms of ontology mismatch, *terminological* and *conceptualization* ontological, which straightforwardly correspond to our definitions of misalignments.

2.2   Concepts Abstraction

The LTS system relies on the concept of *unitary-meaning* or *umeaning*: such atomic concepts can be derived from excerpts from the text of legal norms, such as European directives or national laws, and are arranged into two separate categories of umeanings, as described in [11].

EUDs provide rigorous definitions of some terms, such as the definition of the Italian term *consumatore* (*consumer*), in the Italian version of the *EUD 93/13/EEC*, Art. 2 is:

> [...](b) "consumatore": qualsiasi persona fisica che, nei contratti oggetto della presente direttiva, agisce per fini che non rientrano nel quadro della sua attività professionale; [...]
>
> [...](b) "consumer": means any natural person who, in contracts covered by this Directive, is acting for purposes which are outside his professional activity; [...] (*our literal translation*)

However, two facts must be pointed out. Different EUDs might affect different aspects of the legislation: thus the definition of a term in a EUD only applies to a specific context. Furthermore, EUDs could be written at different points in time, and they can thus introduce diverging definitions. Let us consider the definition of *consumatore*, as it appears in the Italian version of the *EUD 2002/65/EC, Art. 1*:

> [. . .](d) "consumatore": qualunque persona fisica che, nei contratti a distanza, agisca per fini che non rientrano nel quadro della propria attività commerciale o professionale; [. . .]

> [. . .](d) "consumer": means any natural person who, in distance contracts covered by this Directive, is acting for purposes which are outside his business or professional activity; [. . .] (*our literal translation*)

Please note that, in contrast with English, in Italian the second definition of *consumatore* is broader than the first one, since the term *professionale* (*professional*) does not include *commerciale* (*business*). This divergence of term definitions can often occur, since EUDs usually target a sectorspecific sector. In this way, EUDs covering different sectors can provide different definitions, and as many views on the same concept. Lawyers and legislators started to put together highly sector-specific concepts into more abstract concepts with broader meaning, in order to describe (complex) entities, such as the *consumatore* in all of its aspects.

In recent years, in the Italian legislation EUDs are not being implemented as single laws, but rather as groups of EUDs. The juridical concepts are defined as the union of all the sectorial concepts provided by the individual EUDs, as a result of the doctrinal interpretation process of directives.

We remark that these problems are common to all European languages. Consider, for instance the definition of *consumer*, in the English version of the EUD 1999/44/EC, Art. 1.2 is:

> [. . .] (a) consumer: shall mean any natural person who, in the contracts covered by this Directive, is acting for purposes which are not related to his trade, business or profession; [. . .]

that has a different meaning with respect to the definition of *consumer* given in the Council Directive 90/314/EEC, Art. 2.4:

> [. . .] "consumer" means the person who takes or agrees to take the package ('the principal contractor'), or any person on whose behalf the principal contractor agrees to purchase the package ('the other beneficiaries') or any person to whom the principal contractor or any of the other beneficiaries transfers the package ('the transferee') [. . .]

The LTS should be able to represent both the more specific dimension related to the definitions in EUDs and the more abstract one which results from the doctrinal interpretation of European law. The LTS allows inserting the text paragraphs where *umeanings* are defined. However, to gain better understanding of legal concepts, it is often required to consider a broader fragment. For example, in the case of *consumer* the definition is not enough, and it is necessary to collect multiple paragraphs where consumer protection norms are presented and discussed.

2.3   Normative Change

Another big open issue to cope with in building tools for describing legal frameworks, both at the European and at the national level, is the *normative change* [12]. One major problem, well-known in literature, is the update of *non-monotonic* ontologies and knowledge bases [13]. In other words, In other words, ontologies and knowledge bases do not necessarily have a structure which is constant through time (e.g., see [15]): concepts and relations present in the ontology can become obsolete as new concepts and relations are added. This is indeed the case for legal frameworks, that are continuously modified as new laws can modify paragraphs of old ones.

We can have two types of normative change: *explicit* change and *implicit* change. In the first case the new norm explicitly states the abrogation of a specific paragraph of an old law (for details on this line of investigation, please refer to [16,17]). Alternatively, the newer law can state a concept in contradiction with previous laws, but without mentioning these laws explicitly. In this case the concept stated by the new law becomes the current one; also, the parts of the old laws affected by changes are no longer updated and become obsolete.

2.4   Reforming European Law: Toward a Common Frame of Reference

In February 2003, the European Commission adopted a further communication entitled "A More Coherent European Contract Law - An Action Plan" [18]. One of the key measures proposed in the Action Plan was the elaboration of a Common Frame of Reference (CFR). According to the Action Plan, in which the idea of a CFR was developed for the first time, a major aim of the forthcoming CFR is to serve as a tool for the improvement of the EC law. The future CFR was described in more detail in the Commission's Communication on "European Contract Law and the Revision of the Acquis: The Way Forward" [19]. It proposed that the CFR should provide fundamental principles of contract law, definitions of the main relevant abstract legal terms and model rules of contract law. Its main purpose is to serve as a legislators' toolbox.

In drafting the Action Plan the Commission emphasized that the CFR would eliminate market inefficiencies arising from the diverse implementation of European directives, providing a solution to the *non-uniform* interpretation of European contract law due to vague terms and rules, now present in the existing Acquis.

In particular, two issues arise from the vague terminology of EUDs. First, directives adopt broadly defined legal concepts, therefore leaving too much freedom in their implementation to national legislators or judges. Second, directives introduce legal concepts that are different from national legal concepts. Thus, when judges face vague terms, they can either interpret them by referring to the broad principles of the Acquis communautaire, the existing body of EU primary and secondary legislation as well as the European Court of Justice decisions [20], or they can refer to the particular goals of the directive in question. To respond to the Action Plan, in the last few years, within the general framework of a "Network of Excellence" European Project, a research group aiming at consolidating the existing EC law is working on the "Principles of the Existing EC Private Law" or "Acquis Principles" (ACQP). These Principles will be discussed and compared with other outcomes from different European research groups and, during a complex process of consultation with stakeholders under the direction of EC Commission, the CFR will be set up. The Acquis Principles should provide a common terminology as well as common principles to constitute a guideline for uniform implementation and interpretation of European law [20,21]. One outcome of such project is the Acquis Principles glossary, i.e., a set of interconnected terms and concepts.

The Acquis Principles have been sketched by scholars in European Private Law from the Acquis communautaire. Nowadays the corpus contains some 80, 000 pages. Notwithstanding the importance of this existing body of settled laws, the Acquis also has a far wider range, encompassing an impressive set of principles and obligations, going far beyond the internal market and including areas, such as agriculture, environment, energy and transports.

In this paper we show how the multilevel architecture of LTS allows us to relate the Acquis Principles with the legal concepts defined in the directives.

3   The Legal Taxonomy Syllabus

In this Section we first summarize the functionalities of the existing LTS [11], and we then explain how it has been extended to cope with the new requirements described in the previous Section.

3.1   The Basic LTS

The main assumptions of our methodology come from studies in comparative law [7] and ontology engineering [14]:

- Terms –*lexical entries* for legal information–, and concepts must be distinguished; for this purpose we use lightweight ontologies [22], i.e. simple taxonomic structures of primitive or composite terms together with associated definitions. They are hardly axiomatized as the intended meaning of the terms used by the community is more or less known in advance by all members, and the ontology can be limited to those structural relationships among terms that are considered as relevant [23].
- We distinguish the ontology implicitly defined by EUD, the *EU level*, from the various national ontologies. Each one of these "particular" ontologies belongs to the *national level*: i.e., each national legislation refers to a distinct national legal ontology. We do not assume that the transposition of an EUD automatically introduces in a national ontology the same concepts that are present at the EU level.
- Corresponding concepts at the EU level and at the national level can be denoted by different terms in the same national language.

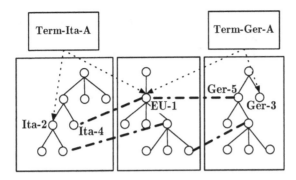

Fig. 1. Relationship between ontologies and terms. The thick arcs represent the inter-ontology "association" link.

A standard way to properly manage large multilingual lexical databases is to make a clear distinction among terms and their interlingual relations (or *axies*) [24,25].

In the LTS project, to properly manage terminological and conceptual misalignment, we distinguish the notion of *legal term* from the notion of *legal concept* and we build a systematic classification based on this distinction. The basic idea in our system is that the conceptual backbone consists of a taxonomy of concepts (ontology) to which the terms can refer in order to express their meaning. One of the main points to keep in mind is that we do not assume the existence of a single taxonomy covering all languages. In fact, the different national systems may organize the concepts in different ways. For instance, the term *contract* corresponds to different concepts in common law and civil law, where it has the meaning of *bargain* and *agreement*, respectively [26]. In most complex instances, there is no synonymy relation between terms-concepts such as *frutto civile* (legal fruit) and *income* from civil law and common law respectively, but these systems can achieve functionally similar operational rules thanks to the functioning of the entire taxonomy of national legal concepts [27]. Consequently, the LTS includes different ontologies, one for each involved national language plus one for the language of EU documents. Each language-specific ontology is related via a set of *association* links to the EU concepts, as shown in Fig. 1.

Although this picture is conform to intuition, in the basic LTS it has been implemented by taking two issues into account. First, it must be observed that the various national ontologies have a reference language. This is not the case for the EU ontology. For instance, a given term in English could refer either to a concept in the UK ontology or to a concept in the EU ontology. In the first case, the term is used for referring to a concept in the national UK legal system, whilst in the second one, it is used to refer to a concept used in the European directives. This is one of the main advantages of LTS. For example *klar und verständlich* could refer both to concept Ger-379 (a concept in the German Ontology) and to concept EU-882 (a concept in the European ontology). This is

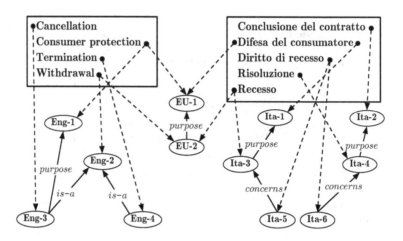

Fig. 2. An example of interconnections among terms

the LTS solution for facing the possibility of a partial correspondence between the meaning of a term in the national system and the meaning of the same term in the translation of a EU directive. This feature enables the LTS to be more precise about what "translation" means. It makes available a way for asserting that two terms are the translation of each other, but just in case those terms have been used in the translation of an EU directive: within LTS, we can talk about direct EU-to-national translations of terms, and about *implicit* national-to-national translations of terms. In other words, we distinguish between *explicit* and *implicit* associations among concepts belonging to different levels. The former ones are direct links that are explicitly used by legal experts to mark a relation between concepts. The latter ones are indirect links: if we start from a concept at a given national level, by following a direct link we reach another concept at European level. Then, we will be able to see how that concept is mapped onto further concepts at the various national levels.

The situation enforced in LTS is depicted in Fig. 1, where it is represented that the Italian term *Term-Ita-A* and the German term *Term-Ger-A* have been used as corresponding terms in the translation of an EU directive, as shown by the fact that both of them refer to the same EU-concept EU-1. In the Italian legal system, *Term-Ita-A* has the meaning Ita-2. In the German legal system, *Term-Ger-A* has the meaning Ger-3. The EU translations of the directive is correct insofar no terms exist in Italian and German that characterize precisely the concept EU-1 in the two languages (i.e., the "associated" concepts Ita-4 and Ger-5 have no corresponding legal terms). A practical example of such a situation is reported in Fig. 2, where we can see that the ontologies include different types of arcs. Beyond the usual *is-a* (linking a category to its supercategory), there are also the arcs *purpose*, which relate a concept to the legal principle motivating it, and *concerns*, which refer to a general relatedness. The dotted arcs represent the reference from terms to concepts. Some terms have links both to a National ontology and to the EU Ontology (in particular, *withdrawal* vs. *recesso* and *difesa del consumatore* vs. *consumer protection*).

The last item above is especially relevant: note that this configuration of arcs specifies that: 1) *withdrawal* and *recesso* have been used as equivalent terms (concept EU-2) in some European Directives (e.g., Directive 90/314/EEC). 2) In that context, the term involved an act having as purpose some kind of protection of the consumer. 3) The terms used for referring to the latter are *consumer protection* in English and *difesa del consumatore* in Italian. 4) In the British legal system, however, not all *withdrawals* have this goal, but only a subtype of them, to which the code refers to as *cancellation* (concept Eng-3). 5) In the Italian legal system, the term *diritto di recesso* is ambiguous, since it can be used with reference either to something concerning the *risoluzione* (concept Ita-4), or to something concerning the *recesso* proper (concept Ita-3).

The LTS is a theoretical instrument as well as a software platform that is operational at the present time. The actual number of annotated terms and concepts are provided in Tables 1 and 2, respectively. Terms were initially extracted from a *corpus* of 24 EC directives, and 2 EC regulations, reported in Appendix 4. Occurrences of such entries were detected from national transposition laws of English, French, Spanish, Italian and German jurisdictions.

Finally, it is possible to use the LTS to translate terms into different national systems via the transposed concepts at the European level, i.e. by using the implicit associations. For instance suppose that we want to translate the legal term *credito al consumo* from Italian to German. In the LTS *credito al consumo* is associated to the national umeaning Ita-175. We find that Ita-175 is the transposition of the European umeaning EU-26 (*contratto di credito*). EU-26 is associated to the German legal term *Kreditvertrag* at European level. Again,

Table 1. Number of terms

Language	National	European
French	8	47
Italian	28	52
English	71	75
Spanish	41	60
German	66	98
total	214	332

Table 2. Number of concepts

Language	National	European
French	7	43
Italian	24	45
English	54	71
Spanish	34	56
German	52	75
total	171	290

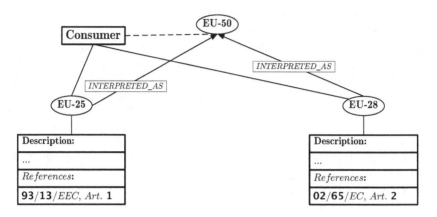

Fig. 3. Umeanings `Eu-25` and `Eu-28` are interpreted by the more abstract umeaning `Eu-50`, the link between `Eu-50` and the term "consumer" is implicit

we find that the national German transposition of `EU-26` corresponds to the national umeaning `Ger-32` that is associated with the national legal term *Darlehensvertrag*. Then, by using implicit links in the European ontology, we can translate the Italian legal term *credito al consumo* into the German legal term *Darlehensvertrag*.

3.2   Enhancing LTS with Interpretation and Abstraction

As described in Section 2.2, different pieces of legislations can bear different definitions of terms. Having different detailed definitions is important during the interpretation of sub-domain specific legal cases, but for the general case it is important to have a view that abstracts from the peculiarities of specific domains.

In order to solve this problem we introduced a new kind of ontologic relation called *INTERPRETED_AS*: it is a non transitive relation where the more general umeaning, that we will call *group leader* represents the abstracted concept that groups the meaning of a number of more specific umeanings, that are the sectorial umeanings defined in the individual EUDs or national laws (see Fig. 3).

We have also introduced a number of constraints and integrity checks to ensure that the semantics of the grouping concept is respected and to improve the usability of the system:

- Each umeaning can belong to a single group.
- A group leader cannot exist without group members.
- When the user searches into the umeaning database, more specific umeanings are excluded from the results unless the user explicitly asks to show them, i.e. only the group leaders are shown in the results.

The need to contextualize concepts to the EUDs defining them leads to the need of more complex instruments to deal with the language of the norms.

A umeaning is defined by the legal texts themselves; this makes clear that the creation of umeaning is quite a long task, because it requires the user to search and read a very large number of documents.

In order to ease this process, we developed a database that contains the full versions of the desired EUDs and national laws. In this way, the user can carry out his task according to the following workflow.

- The user creates a new *umeaning* linked with the term he wants to define.
- He selects relevant citation from legal text; consequently, the browser is redirected to a search page and the main term attached to the umeaning is used as the default query.
- After choosing one of the search results, the full text of the legal document is displayed, with the search terms highlighted.
- Finally, the user selects the text that will go in the citation with the mouse and confirms the insertion in the references database.

3.3   LTS with Normative Change

When a new normative is approved and enacted it can define a number of new umeanings; moreover it can happen that the same law can change a number of old umeanings defined by old laws. In particular, these old umeaning can become obsolete and no longer valid. We are aware of the difficulties concerning the modelling of time in artificial intelligence and in formal ontology creation[2]. Anyway, in LTS we adopted a naive solution in order to manage the simpler situation concerning t. In the LTS it was necessary to delete all old umeanings, causing the loss of all historic information from the database, information that is quite valuable for a better understanding of the evolution of the normative. This problem was resolved by using the same solution adopted for the interpretation and abstraction of the norms (Section 3.2), i.e. empowering LTS with a new ontological relation called *REPLACED_BY*.

When the paragraph of an EUD defining a umeaning has been modified by a new EUD, the new one defines a new umeaning that will replace the old umeaning in the ontology. There will be a relation of type *REPLACED_BY* between the two umeanings, where the child umeaning is replaced by the more general umeaning. Also in this case the new ontological relation has some peculiar characteristics that distinguish it from the usual ontological relations (Fig. 4):

- A *REPLACED_BY* relation brings with it a new data field not present in the other relations: the substitution date.
- When the user performs a search in the umeanings database the replaced ones will not be shown, unless the user asks for a certain past date, thus obtaining a snapshot of the legal ontology that was valid at that particular moment.

---

[2] E.g. see [28] for a general survey and [12] for normative systems.

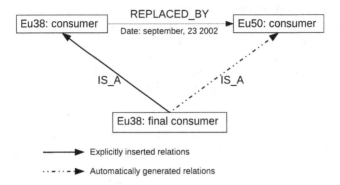

Fig. 4. An example of use of the *REPLACED_BY* relation

- When a new umeaning replaces an old one all the ontological relations in which the old umeaning participated are automatically aplied to the new umeaning. If some of them are no longer valid with the new umeaning, manual intervention from the user is required.

3.4 Enhancing LTS with NLP Techniques

In order to speed up the annotation process, the LTS provides a number of facilities based on natural language processing. In the definition of a new concept it is important to collect all the occurrences of the terms that implement that concept. With this aim, we devised an intelligent search procedure for retrieving the terms which is based on *inverted index, stemming, function words*, and on *document similarity*.

The term search through the legal texts database adopts the standard *inverted index* technique. The inverted index has a quite simple structure: it is a relational table that maps the terms onto the documents containing them, along with the position in the document where terms actually occur. The documents belonging to the national and European levels are stored and indexed in separate relational tables.

In general, the search for a given term in the documents database is performed on the root (*stem*) of the term, instead of searching for the exact term. For instance, when performing the search on the term "contracts", also documents containing only "contract" will be found; this is commonly acknowledged to enhance the information retrieval performance, as shown in [29].[3]

Inverted index based on stems can be computationally hard on a large number of directives. By contrast, the index size can be reduced by excluding the so called

---

[3] The root extraction technique is called *stemming*, and in the LTS we used a library written in the *Snowball* language (http://snowball.tartarus.org), relying on the Porter algorithm [30].

function words (e.g. prepositions, articles), i.e. words with a precise grammatical function but without a real semantic content. In addition, we exclude a number of *common words*, e.g. "mio" ("my") that do not improve the quality of the search. By excluding these lists of word we significantly reduced the number of indexes and improved the performance of the system. For instance, the size of the index of the EUD 90/314/EC decreases from 2336 records to 1152.[4]

By using the inverted index technique we obtain a statistical distribution of the words in the documents as a side effect. Such statistics allow us to define a document similarity measure that can be used into a number of tasks, e.g. to choose the next Directive to annotate with respect to a number of prefixed concepts/terms. The computation of the similarity index between documents relies on the so called *cosine document similarity* technique [32,33].

3.5   Representing a New Perspective in LTS: The Acquis Principles Level

One major feature of the LTS approach relies on distinguishing legal information as belonging to different levels. At the current stage of development, the system manages terms and meanings at both EU and national levels. The former one is an ontology of legal concepts derived from the EUDs; the latter one includes national legal ontologies coming from the various national legal systems. The current approach has been devised to be general enough to account for heterogeneous legal sources (like, e.g., EUDs and "Decreti Legislativi" for European and Italian national levels respectively), and flexible enough to be extended by adding further levels. To add a new level into the system, we connect a new legal ontology to an existing one. The new level is linked via *explicit* associations connecting a concept belonging to the new ontology and a concept belonging to the existing ontology. We are applying this procedure in order to define an *Acquis level* to the LTS.

We introduce the *Acquis level* into the LTS by defining explicit associations between Acquis Principles concepts and EU-level concepts. For example, in Fig. 5 we have that the concept EU-25 (corresponding to the English legal term *creditor*) present in a EUD is explicitly associated with the national legal concepts Ita-124 (*finanziatore*) and Spa-110 (*prestamista*) for Italian and Spanish, respectively. We can add the term *creditor* from the Acquis Level by inserting an explicit association between the Acquis legal concept AC-72 and the European legal concept EU-25. As a consequence, the concept AC-72 is implicitly associated to the legal concepts Ita-124 and Spa-110. This fact has deep consequences on the way one can build systems for reasoning, that are allowed to make paths passing through more than two levels, thereby offering new insights (and ready-to-use associations between terms) to scholars in comparative law.

---

[4] We adopted the list of function words described in [31] (http://members.unine.ch/jacques.savoy/clef/)

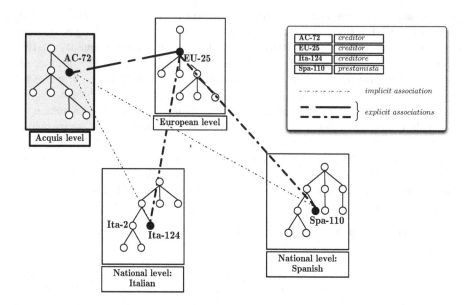

Fig. 5. LTS augmented with the Acquis level. Thick lines indicate *explicit* associations; thin lines indicate *implicit* associations.

4 Conclusions

In this paper we have discussed some new features that have recently been introduced in the LTS, a tool for building multilingual conceptual dictionaries for the EU law. The tool is based on lightweight ontologies to emphasize the distinction between concepts and terms. Different ontologies are built at the EU level and for each national language, to deal with polysemy and terminological and conceptual misalignment.

Many attempts have been made to use ontologies in the legal field (e.g. [2,1], and there are two projects that are strictly related to our approach, i.e. the LOIS and DALOS projects. The LOIS project[5] [4][34] aimed at the (semi-)automatic and manual creation of multilingual legal wordnets in the consumer protection domain. It provided a legal extension to the EuroWordnet architecture [5] and connected its legal ontologies to existing core legal and higher level ontologies. The DALOS project[6] [35], aimed at building an ontological-linguistic resource for the purpose of multilingual EU legislative drafting, and is based on the LOIS database for linguistically reliable national transpositions. Whilst the final goal of LOIS is to support applications concerning information extraction and the DALOS project is realized with the aim to help the legislative process, the LTS we have described in this article is concerned with the access of human experts to the EU documents.

---

[5] http://www.loisproject.org
[6] http://www.dalosproject.eu/

This article illustrates how to distinguish between concepts as they are defined in the text of the directives and concepts representing the doctrinal interpretation of the terms. Moreover, we point out how to deal with normative change by introducing a temporal dimension in ontologies. Finally we show how to add further levels of representation to the EU and national levels. In particular, we described a methodology enabling the insertion of a novel set of principles, called Acquis principles, into the LTS platform. The Acquis principles can be encoded into a number of legal concepts (along with a terminology), and belong to a new level of representation, called the Acquis priciples level. These concepts can be integrated into the LTS by new explicit (and implicit) associations, that connect the Acquis principles level with European and national levels.

Two main problems arise in our approach: the first one is theoretical, and it concerns the issue of evaluating the performance of system with more massive data. We would like to show with some quantitative measure the theoretical adequacy of LTS. Secondly, the amount of work needed to annotate the EUDs with concepts, terms and their transpositions, is huge. Future work will involve exploring ways to extend the LTS ontology and populating it at the various levels by semi-automatic approaches [36].

References

1. Després, S., Szulman, S.: Merging of legal micro-ontologies from european directives. Journal of Artificial Intelligence and Law (February 2007)
2. Casanovas, P., Casellas, N., Tempich, C., Vrandecic, D., Benjamins, R.: OPJK modeling methodology. In: Proceedings of the ICAIL Workshop: LOAIT 2005 (2005)
3. Giguet, E., Luquet, P.: Multilingual lexical database generation from parallel texts in 20 european languages with endogenous resources. In: Proceedings of the COLING/ACL 2006 Main Conference Poster Sessions, Sydney, Australia, pp. 271–278, Association for Computational Linguistics (2006)
4. Tiscornia, D.: The Lois Project: Lexical Ontologies for Legal Information. In: Proceedings of the V Legislative XML Workshop. European Press Academic Publishing (2007)
5. Vossen, P., Peters, W., Gonzalo, J.: Towards a universal index of meaning. In: Proc. ACL 1999 Siglex Workshop (1999)
6. Boer, A., van Engers, T., Winkels, R.: Using ontologies for comparing and harmonizing legislation. In: ICAIL, pp. 60–69 (2003)
7. Rossi, P., Vogel, C.: Terms and concepts; towards a syllabus for european private law. European Review of Private Law (ERPL) 12(2), 293–300 (2004)
8. Ajani, G., Boella, G., Lesmo, L., Mazzei, A., Rossi, P.: Multilingual conceptual dictionaries based on ontologies. In: Proc. of V Legislative XML Workshop, Florence, pp. 1–14. European Press Academic Publishing (2006)
9. Ajani, G., Boella, G., Lesmo, L., Martin, M., Mazzei, A., Rossi, P.: A development tool for multilingual ontology-based conceptual dictionaries. In: Proc. of 5th International Conference on Language Resources and Evaluation, LREC 2006, Genoa, pp. 1–6 (2006)

10. Ajani, G., Boella, G., Lesmo, L., Mazzei, A., Rossi, P.: Multilingual Ontological Analysis of European Directives. In: Proceedings of the 45th Annual Meeting of the Association for Computational Linguistics Companion Volume Proceedings of the Demo and Poster Sessions, Prague, Czech Republic, pp. 21–24. Association for Computational Linguistics (2007)
11. Ajani, G., Boella, G., Lesmo, L., Mazzei, A., Rossi, P.: Terminological and ontological analysis of european directives: multilinguism in law. In: 11th International Conference on Arificial Intelligence and Law (ICAIL), pp. 43–48 (2007)
12. Palmirani, M., Brighi, R.: Time model for managing the dynamic of normative system. Electronic Government, 207–218 (2006)
13. Cadoli, M., Donini, F.M.: A Survey on Knowledge Compilation. AI Communications 10(3-4), 137–150 (1997)
14. Klein, M.: Combining and relating ontologies: an analysis of problems and solutions. In: Workshop on Ontologies and Information Sharing, IJCAI 2001, Seattle, USA (2001)
15. The Gene Ontology Consortium: Gene Ontology: tool for the unification of biology. Nature Genetics 25, 25–29 (2000), http://genetics.nature.com
16. Cherubini, M., Giardiello, G., Marchi, S., Montemagni, S., Spinosa, P., Venturi, G.: NLP-based metadata annotation of textual amendments. In: Proc. of Workshop on Legislative XML 2008, JURIX 2008 (2008)
17. Brighi, R., Lesmo, L., Mazzei, A., Palmirani, M., Radicioni, D.: Towards Semantic Interpretation of Legal Modifications through Deep Syntactic Analysis. In: Jurix 2008: The 21st Annual Conference. Frontiers in Artificial Intelligence and Applications. IOS Press, Amsterdam (2008)
18. European Commission: Communication from the Commission to the European Parliament and the Council - A More Coherent European Contract Law - An Action Plan. COM (2003)
19. European Commission: European Contract Law and the revision of the acquis: the way forward. Communication from the Commission COM (2004) (651 final) (2004)
20. Ajani, G., Schulte-Nölke, H.: The Action Plan on a More Coherent European Contract Law: Response on Behalf of the Acquis Group (2003)
21. Schulze, R.: European Private Law and Existing EC Law. European Review of Private Law, ERPL (2005)
22. Giunchiglia, F., Zaihrayeu, I.: Lightweight Ontologies. Technical Report DIT-07-071, University of Trento, Department of Information and Communication Technology, 38050 Povo – Trento (Italy), Via Sommarive 14 (2007)
23. Oberle, D. (ed.): Semantic Management of Middleware. Springer Science+Business and Media, Heidelberg (2005)
24. Sérasset, G.: Interlingual lexical organization for multilingual lexical databases in NADIA. In: Proc. COLING 1994, pp. 278–282 (1994)
25. Lyding, V., Chiocchetti, E., Sérasset, G., Brunet-Manquat, F.: The LexALP information system: Term bank and corpus for multilingual legal terminology consolidated. In: Proc. of the Workshop on Multilingual Language Resources and Interoperability, ACL 2006, pp. 25–31 (2006)
26. Sacco, R.: Contract. European Review of Private Law 2, 237–240 (1999)
27. Graziadei, M.: Tuttifrutti. In: Birks, P., Pretto, A. (eds.) Themes in Comparative Law. Oxford University Press, Oxford (2004)
28. Allen, J.: Towards a general theory of action and time. Artificial Intelligence 23(2), 123–154 (1984)
29. Krovetz, R.J.: Word sense disambiguation for large text databases. PhD thesis, University of Massachusetts (1995)

30. Porter, M.: An Algorithm for Suffix Stripping. Program 3(14), 130–137 (1980)
31. Fox, C.: A stop list for general text. SIGIR Forum 24(1-2), 19–21 (1990)
32. Lin, D.: An information-theoretic definition of similarity. In: ICML 1998: Proceedings of the Fifteenth International Conference on Machine Learning, pp. 296–304. Morgan Kaufmann Publishers Inc., San Francisco (1998)
33. Isaacs, J.D., Aslam, J.A.: Investigating measures for pairwise document similarity. Technical Report PCS-TR99-357, Dartmouth College, Hanover, NH, USA (1999)
34. Peters, W., Sagri, M., Tiscornia, D.: The Structuring of Legal Knowledge in LOIS. Artficial Intelligence and Law 2(15) (2007)
35. Francesconi, E., Tiscornia, D.: Building Semantic Resources for Legislative Drafting: the DALOS Project. In: Casanovas, P., Sartor, G., Casellas, N., Rubino, R. (eds.) Computable Models of the Law. LNCS (LNAI), vol. 4884, pp. 56–70. Springer, Heidelberg (2008)
36. Cimiano, P.: Ontology Learning and Population from Text: Algorithms, Evaluation and Applications. Springer, Heidelberg (2006)

APPENDIX A: List of EC Directives

Core Directives

- 84/450/EEC concerning misleading advertising
- 85/374/EEC concerning liability for defective products
- 85/577/EEC to protect the consumer in respect of contracts negotiated away from business premises
- 87/102/EEC concerning consumer credit
- 90/88 concerning consumer credit
- 90/314/EEC on package travel, package holidays and package tours
- 93/13/EEC on unfair terms in consumer contracts
- 94/47/EC on the protection of purchasers in respect of certain aspects of contracts relating to the purchase of the right to use immovable properties on a timeshare basis
- 97/7/EC on the protection of consumers in respect of distance contracts
- 97/55/EC concerning misleading advertising so as to include comparative advertising
- 98/6 on consumer protection in the indication of the prices of products offered to consumers
- 98/7 concerning consumer credit
- 98/27/EC on injunctions for the protection of consumers' interests
- 99/44/EC on certain aspects of the sale of consumer goods and associated guarantees
- 2000/13/EC relating to labelling, presentation and advertising of foodstuff
- 2001/95 on general product safety
- 2002/65/EC concerning the distance marketing of consumer financial services
- Regulation 2006/2004/EC on co-operation between national authorities responsible for the enforcement of consumer protection laws
- Directive 2005/29/EC on Unfair Commercial Practices

Ancillary Directives

- 76/768/EEC relating to cosmetic products
- 88/378/EEC on toy safety
- 89/552/EEC on TV broadcasting activities
- 96/74/EC on textile names
- 97/5/EC on cross border credit transfers
- Recommendation 98/257 on the principles applicable to bodies responsible for the out-of-court settlement of consumer disputes
- 2000/31/EC on electronic commerce
- Regulation 2560/2001/EC on cross-border payments in Euro

PART III

Legal Text Processing and Semantic Indexing, Summarization and Translation

Semantic Indexing of Legal Documents

Erich Schweighofer

Centre for Computers and Law
DEICL/AVR, Faculty of Law, University of Vienna
Schottenbastei 10-16/2/5, 1010 Wien, Austria
Erich.Schweighofer@univie.ac.at
rechtsinformatik@univie.ac.at

Abstract. Automated semantic indexing may be the answer to insufficient recall of legal information systems. The semantic web has created powerful tools for mark-up and ontological representation. Re-use in legal applications remains low due to inappropriate knowledge structuring and lack of automated knowledge acquisition. This paper describes the state of the art and proposes a dynamic electronic legal commentary.

Keywords: Legal Ontologies, Legal Retrieval Systems, Semantic Web, Conceptual Indexing.

1 Introduction

With indexing, meta knowledge is added to a knowledge source. The concept is based on Latin origins: *index* ("a discoverer, informer, spy; of things, an indicator, the forefinger, a title, superscription") and *indicare* ("to point out, show"). Structural knowledge is represented or made explicit about knowledge on the world, the nature, people, human beings and activities [1].

In law, indexing is a very important tool for coping with the vast body of legal materials. Firstly, an index of concepts or legal sources is added as an additional entry point to the sequential structure of handbooks, textbooks or collections of materials. Secondly, summaries of cases (head notes) identify the important parts of court decisions. Thirdly, huge reference systems on legal materials are established as navigators in a legal order, either based on citations (e.g. the Austrian index [2] or thesauri (e.g. the Swiss thesaurus [3]).

The medium determines the power of the index. In the Gutenberg era, the paper card was the tool for organizing the conceptual structure but had its limits in quantity and organization. Only a limited number of meaningful concepts could be taken into consideration for this task. Since many years now, databases have taken over providing cheaper and more efficient handling of an index. Quantitative restrictions on the size of the index or its Boolean combinations are gone. Standard information retrieval systems, databases and search engines [4] allow an index of more than 100.000 words without any significant reduction of search time. Automated indexing is standard in the simplified version of indexing with all words except stop words. Internet search engines prove that no limits exist in the size of the text corpus. The inverted index

E. Francesconi et al. (Eds.): Semantic Processing of Legal Texts, LNAI 6036, pp. 157–169, 2010.
© Springer-Verlag Berlin Heidelberg 2010

contains all meaningful words of a document corpus seems the access to knowledge today.

Since the early 1990ies, legal retrieval systems (or legal information systems) have changed enormously. Gone are the days when interface, access, document presentation and search were prone to slow service and occasional error. Internet standards have established the basis that legal retrieval systems have developed into huge text archives of their respective legal systems providing reliable 24 hours/7 days access at low or even no cost. In practice, the text archive and the much more efficient handling of digital documents are the most prominent advantages.

Two methods of legal search can be distinguished: information retrieval (IR) and artificial intelligence (AI). IR in law is finding legal documents in an unstructured text corpus that satisfies the information need. AI in law is finding solutions to legal problems. Semantic indexing means that methodology moves from the dominant IR search (mostly Boolean) to more powerful AI search in the future.

Seen from the perspective of IR, legal search should provide the 20 most relevant documents in a text corpus of millions of documents. Some mark-up (e.g. classifications, thesauri, citations) has always been part of legal retrieval systems [5, 6], but use of these meta data seems to have been diminished due to the prominence of Google search. If only very few keywords are matched with a huge index consisting of a large and partly unknown vocabulary, discrimination is already mathematically difficult and practically impossible. Only iterative search strategies lead to satisfying results. It should be noted that good knowledge of the domain plays a very important role for efficient searching. Already 20 years ago, evaluations of performance on recall and precision of legal information systems were disappointing [7]. Due to bigger text corpora and more complex language, the situation has deteriorated. No study is available so far but the present trend of digital commentaries as "the index" to the text archive is illustrative. Boolean search should find documents already but not precisely known whereas digital commentaries provide access to knowledge.

Semantic web technologies like XML, RDF and OWL provide a platform for the extension of current legal retrieval systems with semantic meta data. Semantic indexing adds the aspect of AI to legal search.

The main concepts of AI are knowledge representation and search for solutions [6]. Both concepts depend heavily on each other: poor knowledge representation requires extensive search; best knowledge representation allows automation of reasoning. The search problem addresses this interrelationship that is central to semantic indexing. The main question is how search can be improved by adding meanings to a text corpus.

Semantics is the study of meaning in communication. "In linguistics, semantics is the subfield that is devoted to the study of meaning, as inherent at the levels of words, phrases, sentences, and even larger units of discourse (referred to as *texts*). The basic area of study is the meaning of signs, and the study of relations between different linguistic units." [8].

The main types of semantic meanings in law are: vocabulary and legal language, legal thesauri, citations, classifications, world ontologies and legal ontologies. A detailed description and analysis of these types will be given below.

As well known in AI applications, semantic indexing faces a serious scaling-up problem if it cannot be properly automated. Tools for semi- and full-automated summarization and indexing are essential for its practical implementation.

The remainder of this paper is organized as follows: Section 2 describes the options and challenges for semantic structuring of legal knowledge, Section 3 gives an overview on knowledge acquisition, semi-automatic summarization and indexing. In Section 4, the idea and present status of a dynamic electronic commentary is presented. Section 5 contains conclusions and future work.

2 Options and Challenges for Semantic Structuring of Legal Knowledge

Semantic structuring in law has some particularities due to the legal domain and its language. The legal text corpus is not inherently structured and a formal taxonomy does not exist. Legal structuring as such is done by lawyers, in their minds, and is presented and made explicit in their argumentations and writings. As a product of this process, a legal commentary is considered as the highest level of this endeavour. The understanding of logic remains also quite different from the formal logic of computer science: its open legal concepts, inherent dynamics of law, system models and syntactic ambiguities provide strong impediments to formalisation.

Thus, a mind gap still exists between traditional legal structuring and formalization. Whereas legal language and conceptual structures are starting points for both methods, the refinement and subsequent knowledge representation and management still differ tremendously. It may be also noted that perceptions on quality do not match either. Legal dogmatics still aims for a most comprehensive and detailed representation of legal authorities. Computer-supported knowledge representation remains in my view still too fascinated by highly advanced structuring and formalization but misses too often the scaling-up requirement in order to be taken seriously by lawyers.

Legal ontologies as an explicit formalization of a domain may bridge this gap between formal logic required for automated legal applications and the classical logic of jurisprudence. They could function as the missing link between the AI & law and the theory of law. The lack of a sufficient number of explicit specifications of knowledge could thus be solved.

The start of our analysis will be legal language that constitutes the most important tool for legal knowledge representation. Thesauri (or legal dictionaries) getting more importance now as a traditional tool for representation of knowledge about legal language use. When thesauri are transformed into computer-useable formalizations with automation options, a first sketchy legal ontology is already created. More advanced representations may formalize complex legal rules and conceptual structures.

2.1 Legal Language

The body of regulatory knowledge must be communicated from international, European, national, regional and communal and other lawmakers to the citizens [9, 10]. Text (in legal language) is still the dominant tool but pictures and multimedia are gaining in importance.

Legal texts contain a technical terminology with four characteristics [11]: Specialized words and phrases unique to law (e.g. *erga omnes,* tort etc.) are highly used. Many quotidian words have different meanings in law, e.g., action (lawsuit) or party (a principal in a lawsuit). Legal writing employs many old words and phrases but also loan words and phrases from other languages (e.g. Latin *prima facie*).

This legal vocabulary contains lexical items with specific semantic meaning due to the conceptual model of a legal domain. Practice of lawmakers and courts has been arbitrarily added to difficulty and complexity. In short, legal vocabularies contain open-textured terms, are inherently dynamic, and the norms in which legal terms are used, are syntactically ambiguous. This allows for contradictions to arise from judicial problem solving.

Legal retrieval systems represent this legal knowledge in an unstructured way containing all synonyms, polysems or homonyms. Legal writers have developed a complex structure of concepts as an abstraction from the text corpus as represented in legal databases. This abstraction shows the (supposed) logical and conceptual structure.

This assessment is confirmed by statistical analysis of legal sources. It is well known that statistical tools for ranking do not sufficiently work in legal databases. On the web or in news databases, statistical word distribution provides support for information filtering and classification. In law, results are interesting but by far not satisfying [12].

2.2 Thesauri and Concept Jurisprudence

A thesaurus for indexing contains a list of every important term in a given domain of knowledge and a set of related terms for each of these terms [13, 14]. A lexical ontology builds up from this basis with works on glossaries and dictionaries, extends the relations and makes this knowledge computer-usable in order to allow intelligent applications. More advanced conceptual work has a long tradition in law. This concept jurisprudence provides definitions and structure to a legal domain. Lexical ontologies add the formalisation method to these representations hat can be understood and re-used by a knowledge system.

2.3 Ontologies

Ontologies [15] constitute an explicit formal specification of a common conceptualization with term hierarchies, relations and attributes that makes it possible to reuse this knowledge for automated applications. The formalization must be on the one hand sufficiently powerful with regard to the knowledge representation, on the other hand it must offer functionalities for automation as well as tools to be produced automatically (see for lexically based ontologies [16]).

In law, two ontologies are required: A world ontology for understanding the facts and a legal domain ontology for structuring the legal knowledge. Any ontology describing the world and its knowledge is regarded as world ontology. The term as such corresponds with facts of a legal case: state of things, actions performed, events, an actual happening in time and space etc. Lawyers have to understand – with the help of experts – as much as possible of the world and its facts in order to handle properly legal governance. Thesauri or lexical ontologies may mix both ontologies but at a later stage a strict differentiation is needed for the purpose of legal subsumption.

2.4 Semantic Web

The semantic web can be considered as an extension to the current web in providing a common framework that allows data to be shared and reused [17]. According to Tim Berners-Lee, the Semantic Web is "not a separate Web but an extension of the current one, in which information is given well-defined meaning, better enabling computers and people to work in cooperation" [18]. Information available on the web is semantically tagged and linked using the technologies of Resource Description Framework (RDF), XML and URIs. This layer model [19] is based on XML (schema, name spaces) that offers a structuring of documents and data at the syntactic level. The next level forms RDF (schema) using the syntax of XML and providing clear rules for the production of meta-data. RDF describes resources by attributes. The RDF attributes are defined as a valid vocabulary by the RDF schema forming also classes and class hierarchies. The next layer may be a logical one, an inference machine (see for ideas of AI & law on the semantic web [20]). In 2004, the W3C has published, besides RDF, the Web Ontology Language (OWL) for the development of sets of terms called ontologies that can be used for supporting advanced Web search, software agents and knowledge management [21].

Besides establishing the framework, the web has so far not much been changed to a semantic representation and offered a broad high-level structuring of knowledge. In law, the semantic web constitutes a tool for representation of domain knowledge but has so far only been used in legal research.

2.5 WordNet Technologies

WordNet is an online lexical reference system that is an initiative of the linguist George Miller. It has been developed and is being maintained by the Cognitive Science Laboratory at Princeton University [22, 23, 24]. Its design is inspired by current psycholinguistic theories of human lexical memory. It encodes conceptual relationships between terms by arranging them in a hierarchical structure. Words (nouns, verbs, adjectives and adverbs) and their short definitions are grouped into synonym sets (synsets), each representing a specific lexical concept. The synsets are linked by a set of different semantic relations (mainly synonymy/antonymy, hyponymy/hyperonymy, meronymy and morphological relations to reduce word forms). WordNet aims at supporting automatic text analysis and AI applications and at providing an intuitively useable enhanced dictionary. The database of the current version 2.0 contains about 150,000 words organized in 115,000 synsets for 200,000 word-sense pairs.

The WordNet technology primarily aims at linguistic support. As concepts are defined with natural language terms, no semantic definitions exist in a formal language. The definitions remain vague from a legal point of view. It is also evident that re-use for automatic reasoning support is limited [25].

The motivation of the EuroWordNet (EWN) [26] was the support of mono- and cross-lingual information retrieval. Based on the Princeton WordNet technology, lexica for eight European languages were developed and connected by an inter-lingual index (ILI) [27]. Within the EWN, the structure of the WordNet was supplemented with additional semantic-lexical relations and three top-level categories. The top level offers 63 semantic distinctions grouped into 3 types of entities. They can be accessed by the ILI and form together the common semantic framework for all European languages. The work on EWN was finished in 1999 but its framework has been continued by the Global WordNet Association, which builds on the results of Princeton WordNet and EWN and provides a worldwide platform for discussing, sharing and interconnecting WordNets. A standard conversion of the Princeton WordNet to RDF/OWL has been developed under the auspices of the W3C [28]. WordNet has been used in the LOIS project [43].

2.6 Cyc

The aim of the Cyc project is to provide automated applications with a knowledge base of formally represented "common sense": real world knowledge that can provide a basis for additional knowledge to be gathered and interpreted automatically [29]. At present, over three million facts and rules have been formally represented in the Cyc knowledge base using CycL, Cyc's formal representation language [30].

The huge potential of the Cyc knowledge is still under experimentation. Applications currently available or in development are integration of heterogeneous databases or intelligent search. In the list of potential applications proposed by the Cyc project, law is not specially mentioned; semantic data mining may be close to the proposed development of an electronic commentary. However, no experiments are reported in this direction so far.

2.7 Legal Ontologies

The motivations for the creation of legal ontologies are evident: common use of knowledge, examination of a knowledge base, knowledge acquisition, representation and reuse of knowledge up to the needs of software engineering [31].

Five types of legal ontologies can be distinguished: representations of legal knowledge, conceptual information retrieval systems, multilingual thesauri, advanced lexical ontologies or interchange formats of documents and knowledge.

Legal knowledge representation remains the most important and challenging task of legal ontologies. After important preliminary work. [32, 33, 34], the frame-based ontology FBO of [35] and [36] as well as the functional ontology FOLaw [37] achieved some prominence. Both were formalized with the description language ONTOLINGUA [15] and represent a rather epistemic approach.

The FBO is conceived as a general and re-usable legal ontology, which offers three classes of model primitives, whereby for each unit a frame structure with all relevant attributes is defined. The types of frames are: norm, action and concept.

The aim of FOLaw is the organization and interconnection of legal knowledge, in particular with regard to the conceptual information retrieval. It contains six basic categories of the legal knowledge: normative knowledge, meta-legal knowledge, world knowledge, responsibility knowledge, reactive knowledge and creative knowledge. FOLaw has been used in follow-up projects. The central difficulty of the FOLaw proved to be the modeling of the "world knowledge". The knowledge gained from FOLaw was used in many projects, in particular E-Court and in the development of a core legal ontology called LRI-Core [38]. The goal of E-Court was the semi-automated multi-lingual information management of various sources (audio, video, text) in the field of penal law. LRI-Core is a broad concept structure with typical main legal concepts. Anchors as links between the foundational (upper) ontology (the world knowledge) and the legal core ontology (the legal concepts) would support legal subsumption. E-Power [39] was a project of the Dutch Tax and Customs Administration. Laws and regulations were formalized as conceptual models offering automated tasks (e.g. subsumption, calculation or document assembly) and providing comprehensive support from legislation to application.

An impressive standard for the development of a legal ontology exists now with the LKIF Core Ontology (Legal Knowledge Interchange Format) [40]. This ontology was developed in the Estrella project. An application exists in the field of traffic law. LKIF contains a Standard OWL ontology with OWL-DL (description logic) and description logic programs (DLP). Formalizations of obligations, permissions, roles, rights, duties, privileges, liabilities etc., top level clusters, mereological relations, location, time, changes (processes), agents + actions + roles, propositions, legal agents + actions, rights, powers, norms etc. have been developed. LKIF rules are more expressive than those in OWL. The high quality of LKIF is accepted, however, the number of follow-up applications remains quite low.

Much easier for practical applications are conceptual information retrieval systems. Knowledge is reformulated using legal ontologies allowing advanced and intelligent search. In the CLIME/MILE project [41], a legal information server has been developed for the classification of ships and maritime law. Iuriservice [42] is a web-based decision support for Spanish judges in their first appointment. It consists of a database of FAQ, ontological description of documents, a question topic ontology (QTO), an Ontology of Professional Judical Knowledge (OPJK) and semantic distance calculation for improved retrieval.

Lexical ontologies on multilingual thesauri use the ontological structuring for access to multilingual text corpuses. In the LOIS (Lexical Ontologies for legal Information Serving) project, a tool for multi-lingual access to European legal databases was created. 5000 legal concepts were formally represented in all languages on the basis of the WorldNet technology in six languages (synsets of the ILI inter-lingual index containing also legal definitions) [43]. The Legal Taxonomy Syllabus [44] is a tool for annotation and recovery of multi-lingua legal information on EU directives. It contains legal dictionaries and taxanomies of legal concepts. In the follow-up project of LOIS, the DALOS project, an ontological-linguistic resource for multilingual drafting process in EU was created [45]. The ontological layer consists of a conceptual modelling at a language-independent level. The lexical layer represents lexical manifestations in different languages. The ontology is created using also term extraction with NLP tools.

Advanced lexical ontologies differ from lexical ontologies in the quality of knowledge representation and its use. Projects results of LOIS, Legal Taxonomy Syllabus, Juriservice or DALOS provide the basis. According to the outline of the Core Legal Ontology (CLO), the advanced lexical ontology consists of a world ontology (world knowledge represented in thesaurus entries) and a legal ontology (legal system represented in materials rules, procedural rules and concepts). In both ontologies, concepts are represented as frames. These frames allow for establishing extensive links between the world ontology and the legal ontology providing some support for legal subsumption [46].

Interchange formats for documents and knowledge provide basic support for the development of ontologies. Many interchange standards for documents exist now in international, European and national applications (e.g. E-Law in Austria). With MetaLex, a generic and extensible framework for XML-encoding of legal resources has been developed [47].

3   Knowledge Acquisition, Semi-automatic Summarization and Indexing

The knowledge acquisition task for a LKIF-like application remains to be solved. Quality expectations on knowledge representation in law are very high. Lawyers are liable for any error in their legal work; thus, poor knowledge management cannot be accepted. Knowledge products must be accurate, reliable and up to date on relevant legal authorities.

Lawyers as human knowledge engineers can produce the required quality. The knowledge teams have grown over the years in order to cope with work load and speed. So far, these resources only go to traditional products like handbooks or commentaries. Advanced tools of knowledge acquisition are not used much so far but this situation may change in the next years.

Automation of knowledge acquisition is the next option. However, computational linguistics and natural language processing do not achieve the required quality so far. Therefore, a full automation of knowledge acquisition from a huge text corpus will still remain wishful thinking for a while.

Since the 1990ies, semi-automatic knowledge acquisition, summarization and indexing of legal documents are considered as the most promising option. These methods demand an efficient handling of three components: text corpus, meta data and matching facility.

The text corpus can remain in simple text or HTML or upgraded to a higher representation. In particular, part-of-speech tagging, e.g. mapping the words in a text corpus as corresponding to a particular word class and its relations with adjacent and related words in its context by efficient parsers, constitutes a major improvement.

The meta data consists of a knowledge base with a lexical ontology or higher ontological representations. Such a specialized ontology differs from a standard ontology in the added information for the automated matching task. In particular, information on context, special document parts, or even catch phrases may be added.

The matching facility is based on Boolean search for identical or similar expressions. Simple lexical ontologies are reduced to Boolean search for terms and their synonyms in text corpora. Higher ontologies may take advantage of higher representations with part-of-speech taggers. Quite many prototypes already exist in the related field of semi-automatic text analysis and conceptual indexing (see e.g. the projects KONTERM [6, 12], SALOMON [48], FLEXICON [49], SMILE [50], or Support Vector Machines [51]). Summarization work relies also on automatic corpora-based analysis for finding of concepts.

The automated linking of documents constitutes the most advanced work in semantic indexing (e.g. AustLII [52], CiteSeer [53]). It has to be noted that the task is easier due to more formalized language and a controlled vocabulary.

4 Semantic Indexing and the Dynamic Electronic Legal Commentary

Handbooks or commentaries are the most advanced form of traditional explicit knowledge representation. A legal handbook or commentary consists of two parts that are closely interrelated: a systematically structured analysis of a legal domain and a reference section containing all important documents of relevant authorities (parliaments, courts, administrations, legal authors etc.). The sequential structure can be based on a particular legal act (e.g. in sections or articles *("Kommentar")*) or a conceptual structure *("System")*. These representations are also classified according to the size and depth of analysis.

Legal authors are producing handbooks or commentaries that are thus representing intellectual analysis of a particular knowledge domain. Human brains grasp as much knowledge as possible by reading and studying the factual and legal materials and related legal practice. Handbooks or commentaries are always slightly outdated as it takes some time to analyse the domain. Printing obstacles for regular updates have been overcome by digital editions of commentaries (e.g. the Austrian legal databases RDB [54] and LexisNexis [55]).

A dynamic electronic legal commentary has the same aim - a sufficiently detailed representation of a legal domain – but a very different knowledge representation. The most important tool is not the human brain but the computer. The input – a text corpus of a legal system – is described, indexed, reformulated, extracted, classified, summarized and analyzed by means of semantic indexing. The output is a compressed knowledge representation similar to the traditional commentary. Obviously, the representation cannot be in prose but consists of a highly structured set of hyperlinks to legal materials and world knowledge, semantic indexing of materials as well as a summarisation of most important documents. It will be still the work of legal authors to reformulate this representation into a nicely written legal text if necessary. However, it has to be noted that today large sections of legal commentaries are also mostly references with short hints to contextual content. The main advantage lies in the automation of conceptual indexing. A daily consideration of updates, e.g. a dynamic adaptation, will be possible. It has to be stressed too that sufficient quality of a dynamic electronic legal commentary is subject to the knowledge engineering team. This group of human brains has to develop and maintain the knowledge base but also fulfil regular checks and improvements of its output.

Text corpora with sufficient coverage are available in legal databases. It is mostly a question of the knowledge representation if part-of-speech parsers or a reformulation of text in an abstract representation as a conceptual and logical structure is required or not. The main problem consists in developing appropriate legal ontologies as knowledge representations.

Available legal ontologies are either broad and shallow (e.g. LOIS and DALOS) or small and deep (e.g. LRI Core). Further, world ontologies like WordNet or Cyc must still be improved. Small experiments with the LOIS lexical ontology have shown that the number of 5000 lexical entries was not sufficiently big for a sufficient granularity of the analysis. Thus, a first application of a dynamic electronic legal commentary requires improved ontologies that are presently developed in smaller domains of legal informatics, state aid law and consular and diplomatic protection law at our Centre.

Limited but important options exist for the improvement of legal information systems. Existing or improved mark-up and semantic indexing may be used for tools like a navigator (computation of time layers of the legal order, provision of consolidated texts etc.), a citator (automatic linking of documents) or a terminologist (conceptual analysis of a text corpus). Such instruments can be implemented in short time and would be highly desirable [56]. Prototypes and practical applications already exist, however, automation and semantic indexing can still be improved.

5   Conclusions

Legal semantic indexing and the semantic web share a common fate. Whereas powerful formalizations exist in the form of MetaLex, LKIF or LOIS, legal practice has not used much of it for applications. Advantages of re-use of formalized knowledge and dynamic updating do not seem sufficiently convincing for lawyers. Research is ongoing and the concept of a dynamic electronic legal commentary may be implemented in the near future given sufficiently powerful world and legal ontologies. In the near future, automated tools for the improvement of legal information systems like navigator, citator or terminologist may provide a highly helpful support for users.

References

1. Wikitionary, index (2009), http://en.wiktionary.org/wiki/index
2. Index 2006, Rechtsprechung und Schrifttum, Jahresübersicht 2006. Band 59, Begründet von Franz Hohenecker. Manz, Wien (2007)
3. Jurivoc. Dreisprachiger Thesaurus des Schweizerischen Bundesgerichts (2009), http://www.bger.ch/de/index/juridiction/ jurisdiction-inherit-template/jurisdiction-jurivoc-home.htm
4. Manning, C.D., Raghavan, P., Schütze, H.: Introduction to Information Retrieval. Cambridge University Press, Cambridge (2008)
5. Bing, J. (ed.): Handbook of Legal Information Retrieval. North-Holland, Amsterdam (1984)
6. Schweighofer, E.: Legal Knowledge Representation, Automatic Text Analysis in Public International and European Law. Kluwer Law International, The Hague (1999)
7. Blair, D.C., Maron, M.E.: An Evaluation of Retrieval Effectiveness for a Full-text Document-retrieval System. Comm. ACM 28, 289–299 (1985)
8. Wikipedia, semantic (2009), http://en.wikipedia.org/wiki/Semantic
9. Rathert, M.: Sprache und Recht. Universitätsverlag Winter, Heidelberg (2006)
10. Tiscornia, D.: The Lois Project: Lexical Ontologies for Legal Information Sharing. In: Biagioli, C., Francesconi, E., Sator, G. (eds.) Proceedings of the V. Legislative XML Workshop. European Press Academic Publishing, Firence (2006)
11. Wikipedia, legal writing (2009), http://en.wikipedia.org/wiki/legal_writing

12. Schweighofer, E., et al.: Improvement of Vector Representation of Legal Documents with Legal Ontologies. In: Proceedings of the 5th BIS. Poznan University of Economics Press, Poznan (2002)
13. Wikipedia, thesaurus (2009), http://en.wikipedia.org/wiki/Thesaurus
14. ISO: Documentation – Guidelines for the establishment and development of monolingual thesauri, ISO 2788 (1986)
15. Gruber, T.R.: A Translation Approach to Portable Ontology Specifications. Knowledge Acquisition 5/2, 199–220 (1993)
16. Hirst, G.: Ontology and the Lexicon. In: Staab, S., Studer, R. (eds.) Handbook on Ontologies, pp. 210–229. Springer, Heidelberg (2004)
17. W3C: Semantic Web Activity, http://www.w3.org/2001/sw/
18. Berners-Lee, T., et al.: The Semantic Web. Scientific American 284(5), 34–53 (2001)
19. Koivunen, M.-R., Miller, E.: W3C Semantic Web Activity. In: Proceedings of the Semantic Web Kick-off Seminar, HIIT Publications 2002/1, Helsinki, pp. 27–43 (2002), http://www.w3.org/2001/12/semweb-fin/w3csw
20. Benjamins, R., et al.: Law and the Semantic Web, an Introduction. In: Benjamins, V.R., Casanovas, P., Breuker, J., Gangemi, A., et al. (eds.) Law and the Semantic Web. LNCS (LNAI), vol. 3369, pp. 1–17. Springer, Heidelberg (2005)
21. W3C: OWL Web Ontology Language Semantics and Abstract Syntax, W3C Recommendation (February 10, 2004), http://www.w3.org/TR/2004/REC-owl-semantics-20040210/
22. Miller, G.A., et al.: Five Papers on WordNet, CSL Report 43. Cognitive Science Laboratory, Princeton University (1990), ftp://ftp.cogsci.princeton.edu/pub/wordnet/5papers.ps
23. Fellbaum, C. (ed.): WordNet: An Electronic Lexical Database. MIT Press, Cambridge (1998)
24. WordNet website, http://wordnet.princeton.edu/
25. Fensel, D.: Ontologies: A Silver Bullet for Knowledge Management and electronic Commerce, 2nd edn. Springer, Berlin (2004)
26. EuroWordNet website, http://www.illc.uva.nl/EuroWordNet/docs.html
27. Vossen, P. (ed.): EuroWordNet General Document (LE2-4003, LE4-8328). Final Document, Version 3 (1993), http://www.illc.uva.nl/EuroWordNet/docs.html
28. W3C: RDF/OWL Representation of WordNetW3C Working Draft (June 19, 2006), http://www.w3.org/TR/2006/WD-wordnet-rdf-20060619/
29. Lenat, D.B.: Cyc: a Large-Scale Investment in Knowledge Infrastructure. Communications of the ACM 38(11), 33–38 (1995)
30. Cyc website, http://www.cyc.com/cyc
31. Bench-Capon, T.J.M., Visser, P.R.S.: Ontologies in Legal Information Systems: The Need for Explicit Specifications of Domain Conceptualisations. In: Proceedings of the 6th ICAIL, pp. 132–141. ACM Press, New York (1997)
32. McCarty, L.T.: A Language for Legal Discourse: I. Basic Features. In: Proceedings of the 2nd ICAIL, pp. 180–189. ACM Press, New York (1989)
33. Hafner, C.D.: An Information Retrieval System Based on a Computer Model of Legal Knowledge. UNI Research Press, Ann Arbor (1977)
34. Stamper, R.K.: The Role of Semantics in Legal Expert Systems and Legal Reasoning. Ratio Juris 4/2, 219–244 (1991)
35. van Kralingen, R.W.: Frame-based Conceptual Models of Staute Law. Ph.D. Thesis, University of Leiden, The Hague (1995)

36. Visser, P.R.S.: Knowledge Specification for Multiple Legal Tasks: A Case Study of the Interaction Problem in the Legal Domain. Computer Law Series, vol. 17. Kluwer Law International, The Hague (1995)
37. Valente, A.: Legal knowledge engineering: A modelling approach. IOS Press, Amsterdam (1995)
38. Breuker, J., Hoekstra, R.: DIRECT: Ontology-based Discovery of Responsibility and Causality in Legal Case Descriptions. In: Proceedings of the 17th JURIX. IOS Press, Amsterdam (2004)
39. Van Engers, T.M., Gerrits, R., Boekenoogen, M., Glassée, E., Kordelaar, P.: POWER: Using UML/OCL for modeling legislation - an application report. In: Proceedings of the 8th international conference on Artificial intelligence and law, pp. 157–167. ACM Press, New York (2001)
40. Hoekstra, R., Breuker, J., De Bello, M., Boer, A.: The LKIF Core Ontology of Basic Legal Concepts. In: Casanovas, P., Biasiotti, M.A., Francesconi, E., Sagri, M.T. (eds.) Proceedings of LOAIT 2007, II. Workshop on Legal Ontologies and Artificial Intelligence Techniques, pp. 43–64 (2007),
    http://www.ittig.cnr.it/loait/LOAIT07-Proceedings.pdf
41. Winkels, R., et al.: CLIME: lessons learned in legal information serving. In: Proceedings of the 15th ECAI, pp. 230–234. IOS Press, Amsterdam (2002)
42. Casellas, N., Casanovas, P., Vallbé, J.-J., Poblet, M., Blázquez, M., Contreras, J., López-Cobo, J.-M., Benjamins, R.: Semantic Enhancement for Legal Information Retrieval: IUR-ISERVICE performance. In: Eleventh International Conference on Artificial Intelligence and Law, pp. 49–57. ACM Press, New York (2007)
43. Dini, L., Liebwald, D., Mommers, L., Peters, W., Schweighofer, E., Voermans, W.: LOIS Cross-lingual Legal Information Retrieval Using a WordNet Architecture. In: Proc. Tenth Int. Conf. on Artificial Intelligence & Law, pp. 163–167. ACM Press, New York (2005)
44. Ajani, G., Lesmo, L., Boella, G., Mazzei, A., Rossi, P.: Terminological and Ontological Analysis of European Directives: multilinguism in Law. In: Eleventh International Conference on Artificial Intelligence and Law, pp. 43–48. ACM Press, New York (2007)
45. Francesconi, E., Spinosa, P., Tiscorina, D.: A linguistic-ontological support for multilingual legislative drafting: the DALOS Project. In: Casanovas, P., Biasiotti, M.A., Francesconi, E., Sagri, M.T. (eds.) Proceedings of LOAIT 2007, II. Workshop on Legal Ontologies and Artificial Intelligence Techniques, pp. 103–112 (2007),
    http://www.ittig.cnr.it/loait/LOAIT07-Proceedings.pdf
46. Schweighofer, E.: Computing Law: From Legal Information Systems to Dynamic Legal Electronic Commentaries. In: Magnusson Sjöberg, C., Wahlgren, P. (eds.) Festskrift till Peter Seipel, pp. 569–588, Norsteds Juridik AB, Stockholm (2006)
47. Boer, A., Winkels, R., Vitali, F.: Proposed XML Standards for Law: MetaLex and LKIF. In: Lodder, A.R., Mommers, L. (eds.) Legal Knowledge and Information Systems, JURIX 2007: The Twentieth Annual Conference, pp. 19–28. IOS Press, Amsterdam (2007)
48. Moens, M.-F., et al.: Abstracting of Legal Cases: The SALOMON Experience. In: Proceedings of the 6th ICAIL, pp. 114–122. ACM Press, New York (1997)
49. Smith, J.C., et al.: Artificial Intelligence and Legal Discourse: The Flexlaw Legal Text Management System. Artificial Intelligence and Law 3(1-2), 55–95 (1995)
50. Brüninghaus, S., Ashley, K.D.: Improving the Representation of Legal Case Texts with Information extraction Methods. In: Proceedings of the 8th ICAIL, pp. 42–51. ACM Press, New York (2001)
51. Gonçalves, T., Quaresma, P.: Is linguistic information relevant for the classification of legal texts? In: Proceedings of the 10th ICAIL, pp. 168–176. ACM Press, New York (1995)

52. AustLII website, http://www.austlii.edu.au
53. CiteSeer website, http://citeseer.ist.psu.edu/cs
54. RDB Rechtsdatenbank website, http://rdb.at
55. Lexis Nexis Orac ARD website, http://www.lexisnexis.at
56. Schweighofer, E.: EUR-Lex: from data structures to legal ontologies. In: 25 Years of European Law Online, The Event, pp. 137–150. Office for Publications, Luxembourg (2007)

Automated Classification of Norms in Sources of Law

Emile de Maat and Radboud Winkels

Leibniz Center for Law, University of Amsterdam
{demaat,winkels}@uva.nl

Abstract. The research described here attempts to achieve automated support for modelling sources of law for legal knowledge based systems and services. Many existing systems use models that do not reflect the entire law, and simplify parts of the text. These models are difficult to validate, maintain and re-use. We propose to create an intermediate model that has an isomorphic representation of the structure of the original text. A first step towards automated modelling is the detection and classification of provisions in sources of law. A list of different categories of norms and provisions that are used in Dutch legal texts is presented. These categories can be identified by the use of typical text patterns. Next, the results of experiments in automated classification of provisions using these patterns are presented. 91% of 592 sentences in fifteen different Dutch laws were classified correctly. Some conclusions about the generality of the approach are drawn and future research is outlined.

Keywords: Categorisation of Norms, Experimental Results, Natural Language Processing.

1 Introduction

If we want to design and build systems to support users in handling legal knowledge or data, we will always have to start with the sources of law. After all, in a modern constitutional state, all legal action is grounded in and justified by these sources. Legal texts, however, are meant to be read by humans, and are written in natural language. In order to make these sources available to machines, they need to be translated from natural languages to formal languages. This is a time and effort consuming task, usually performed by knowledge engineers with the aid of legal experts. People are researching ways to support and partially automate this task. A first and relatively easy step is to transform unstructured or badly structured text into a well structured one. We use the MetaLex XML interchange format for legal sources for that.[1] Secondly, we want to find and resolve all references in the text, both internal and to external sources, and tag these explicitly, also using MetaLex. That research has been described before; see e.g. [1]. This chapter will discuss the next step: Recognizing and classifying norms or provisions in the legal sources. The idea is that this classification facilitates the suggestion of model fragments to be used for representing the meaning

---

[1] www.metalex.eu

E. Francesconi et al. (Eds.): Semantic Processing of Legal Texts, LNAI 6036, pp. 170–191, 2010.
© Springer-Verlag Berlin Heidelberg 2010

of these norms and provisions. In the (E)-Power project this classification was left implicit and the step from the surface structure of sentences to a formal representation was typically too large to yield useful automatic translations [2]. Making this intermediate step explicit should bridge this gap to some extent. We will first present a classification of norms at a very general level, based on their function in the legal system (Section 2). In Section 4 we will present a finer grained classification and typical examples from Dutch law. These examples also show the typical language structures legislative drafters use (at least in the Netherlands) that will help us in automatically recognising these norms in legal texts. Next, we will show the results of a classifier based upon these typical language structures (Sections 5 and 6). We will end with a discussion and conclusions (Section 7).

2   Models of Legislative Texts

The goal of a law is (or perhaps: should be?) to set rules for the people (and organisations) living in a country (or whatever the jurisdiction of the law is). It tells them what they can do and cannot do, and what their rights and duties are. So, we could expect a law to consist mainly of statements like "Everybody has the right to freedom of speech" and "If you take care of a child less than eighteen years of age, you have a right to child benefit". This turns out not to be the case. Hart [3] distinguishes two types of rules in a law: primary and secondary rules. The primary rules are the rules that refer to human behaviour. Secondary rules are actually rules about primary rules, and form a meta-level. Three types of secondary rules are given by Hart: rules of recognition, rules of change and rules of adjudication. Rules of recognition determine which rules are 'official', rules of change allow for the changing of rules and rules of adjudication empower individuals to judge whether a rule has been broken.

An example of a primary rule is the following one:

General Child Benefit Law, article 7, sub 1
Conform the stipulations of this law, the insured has a right to child benefit for an own child, a stepchild and a foster child which:
a. is younger than 16 years of age and belongs to his household; or
b. is younger than 18 years of age and is maintained by him for a significant amount.

This sentence gives the right to child benefit to the insured, provided certain conditions are met. On the other hand, the following norm is a typical example of a secondary rule. Although it comes from the General Child Benefit Law, there is nothing in this sentence that helps a citizen determine whether or not he has a right to Child Benefit, and, if so, to what amount of Child Benefit.

General Child Benefit Law, article 24b
By Ministerial Decree additional rules can be set regarding the articles 24, sub 1, 2, 3, 4, 5 and 6, and 24a.

In a sense, these secondary rules can be seen as overhead in the law. We want the law to regulate certain things (in this case, child benefit). In addition to the rules on child benefit, however, we need some rules to fit these "core rules" in the legal framework. Other examples are auxiliary provisions that handle enactment of the law, changes, etc.

It is not only the secondary norms, however, that form overhead. The following sentence, for example, is clearly a primary norm, directing some part of the behaviour of citizens:

General Child Benefit Law, article 14, sub 2
A request is made by means of an application form, which is provided by the Social Insurance Bank.

It is not a "core rule", however, as it does not post a direct rule regarding who receives child benefit and how much. Of course, it is rather obvious why it is present in the law. Unfortunately, by merely specifying who receives child benefit and how much, the benefits are not automatically distributed. A system needs to be set up for this to happen. This leads to two layers of additional procedural overhead. The first layer (to which the example above belongs) is still directed to the behaviour of citizens (or citizens' organisations). These rules are not primary rules, even though they are dealing with citizens' behaviour, as they only exist to support the primary rules. On the other hand, they are not rules of recognition, change or adjudication, and therefore, are not instances of Hart's secondary rules. The second layer is about the internal workings of the government and the duties of civil servants. This second layer will contain similar procedural rules, but this time aimed at civil servants. It will also contain norms of competence (as defined in [4]), as well as Hart's rules of adjudication.

An example of a sentence from this second layer is the following:

General Child Benefit Act, article 17d, sub 3
The Attorney General will inform the Social Insurance Bank of any circumstances as meant under sub 1 or 2.

All in all, we come to a four-layer model of the law, as illustrated in Figure 1:

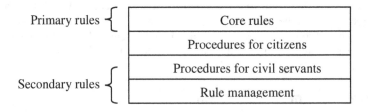

Fig. 1. Layers of norms

This model depicts the four layers of norms as we have described above. For a more complete model of the law, we must remind ourselves that not all the text in a law forms actual norms. In addition to the norms, the body of the law contains definitions. Although definitions could be considered to be normative (for example, a definition of a car could be considered a normative statement determining what may or may not be called a car for the purpose of this law), it is certainly not a normative rule in the sense of e.g. Hart, as it does not deal with human behaviour (primary rule), nor with other rules (secondary rule). The difference with normative rules becomes even clearer when the sentences are studied without context: the rules will usually still have some meaning, though without their accompanying definitions, they are probably vaguer. The definitions, on the other hand, have no meaning outside of the law.

Most primary rules will make use of definitions, but also the procedural rules introduced above. Thus, we cannot say that definitions belong to one specific level of the law. Together with rules, the definitions make up the body of the law. Add to that the introduction, conclusion and appendices, and we come to a more complete model of a legislative text:

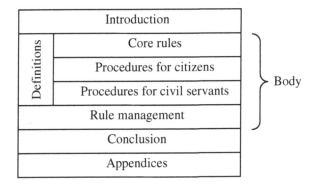

Fig. 2. More complete model of a law text

Most Knowledge Management Systems will focus on the body (and perhaps appendices) of a law text, as the introduction and conclusion will generally not include the kind of information needed for these systems. It is interesting to note that most systems do not focus on all the layers we have distinguished. Most systems that are aimed at given information to citizens will be aimed at the core rules and the procedures for citizens (for example our system for the Legal Services Counter [5]), whereas a process management system for civil servants will most likely incorporate only the two procedural layers. Systems for searching through laws will only model the secondary rules (if any), as those rules are the ones that link the laws together (see for example the Tax Administration Semantic Network [6]).

There have been few attempts to model a law in its entirety. Even within the POWER project [2], which was aimed at laws in their entirety, secondary rules were never modelled and procedural norms often left out. Furthermore, when a model of a law is made, it is usually a model of the meaning of the law, not of the law as a text. This means that though the model will generate the correct outcome, it will not always do so based on the same structure followed by the legal text.

A third remark regarding the existing models is that they are often targeted to a specific population, and therefore simplified to fit that population. For example, when there is a rule that applies to people younger than 25 or older than 65, the second part will often be omitted in a model for an application aimed at young people. Terms will also often be simplified or (partially) interpreted. For example, the law may use the word "vehicle" which may be simplified to "car" in a specific model.

These models do bring potential problems when they need to be updated (because of a change in the legislation) or when one wants to re-use the model. Because of the simplifications, the model does no longer have a one-to-one correspondence to the sentences in the law, which makes it more difficult to determine which elements of the model are affected by a textual change in the law. Similarly, unless they have been clearly documented, any interpretations are difficult to undo or modify.

To avoid such problems, it would be preferable to first create a model of the entire law, and then derive specific application models from it, as depicted in Figure 3.

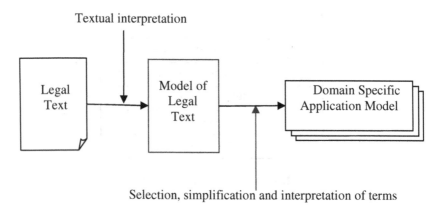

Fig. 3. Model of legal text and domain specific application model

This intermediate model of the law has a clear correspondence to the original text, adhering to the isomorphic representation principle [7]. It should be distinguished from the MetaLex version of the law that deals with structure and referencing. Based on this structure, individual norms or rules are identified (see below) and this leads to suggested model fragments (types of norms are associated with typical representation patterns). It also differs significantly from the integrated, simplified or specialized application specific model that should be built afterwards.

Such an intermediate model should be easy to maintain, update and re-use. Ideally, such a model would be published by the legislator, next to the original textual version of the law, using some generic format, such as perhaps LKIF [8].

3   Granularity for Finding Norms

To make a model of an entire law, each element of the law text should be represented in the model. What is the level of granularity that we should use? Input to the process is a structured, MetaLex version of the law. MetaLex distinguishes and identifies articles, full sentences and sentence fragments. The leading principle in MetaLex design was that the smallest unit that could be referred to should be identifiable. Legal texts can have references of the form "the last sub clause of the previous sentence".

Typically for law texts, articles are the units that can be read and understood without the context of the rest of the law (this does not imply that we do not need the rest for 'correct' application of the law of course). Within articles, sentences are the smallest units that are meaningful by themselves. For our purpose we will start at the sentence level. We assume that each sentence contains a single norm (from which an inverse norm may be derived[2]). Our current research shows some other possibilities in Dutch legal text.

First of all, it may be the case that a sentence contains two main sentences, as in the following example:

> **Bailiff Law, article 3a, sub 3, second sentence**
> Because of required speed, the notice may be given orally, in which case it is immediately confirmed in writing.

This sentence describes both the right to give the notice orally, as the duty to confirm the notice in writing (in those cases).

Another situation in which a sentence can contain multiple norms arises when a subordinate clause contains an (implicit) norm:

> **General Child Benefit Law, article 14, sub 2**
> A request is made by means of an application form, which is provided by the Social Insurance Bank.

This sentence contains the (explicit) duty to use specific forms to make a request, and an (implicit) obligation for the Social Insurance Bank to provide such a form.

A third category is formed by sentences like this one:

> **Customs Law, article 13, sub 1**
> During the investigation, as meant in article 12, only civil servants of the Tax Administration, that have been selected by the inspector, may enter a house without the inhabitant's permission.

In essence, this sentence contains a permission: *Civil servants of the Tax Administration that have been selected by the inspector may enter a house without the inhabitant's*

---

[2] There are always two sides to a norm. For example, if A is obliged to pay a certain amount of money to B, then B has the right to receive that amount from A. A legal text will only mention one of those two norms; the other can be derived.

permission. Adding the word "only" adds a prohibition to the sentence: it is forbidden for any other person to enter the house. Without the word "only" the sentence does not say anything about other persons.

4 Sentence Types and Patterns

As explained in Section 3, we treat the sentence as the smallest unit that contains a single norm. Obviously, different sentences in a legal source will lead to different models. Therefore, it is important to distinguish between the different types of sentences that appear in legal sources. In this section, we will present a number of sentence types that can be distinguished. Biagioli et al. [9] have used similar categories. However, their categories were based on whole provisions (articles) instead of single sentences, and classified normative provisions as a permission, duty or exception. Sentences on the other hand can be classified in a more detailed way than provisions, so we have some additional classes as will be shown below.

This categorisation is based on earlier research [10] into the types of sentences that occur in laws. However, this earlier research was largely based on a single (though extensive) law, the Income Tax Act 2001. In addition, as it was part of the (E-)POWER project, it focused on the core rules and procedures expressed in the law, and did not pay attention to the rule management part. Hence, the patterns needed to be extended. About 20 other Dutch laws were studied for this purpose, mostly from the last 50 years, but one going back to 1875.

4.1 Definitions

As mentioned in Section 2, definitions are used to describe the terms that occur in a legal source. The terms defined can later be used in both primary and secondary norms. A definition mentions both the term being defined as well as the actual definition. An example of a definition is:

| **General Administrative Law Act, article 1:4, sub 1** |
| By administrative judge is understood: an impartial body that is appointed by law and charged with administrative judicial settlement |

In most Dutch legal texts, definitions use clear patterns, such as "*onder ... wordt verstaan*" (by ... is understood). This gives us a good method of recognising definitions.

Type extensions are very similar to definitions. However, instead of completely defining a new term, they expand or limit an earlier definition. The most common use of type extensions is to expand a *common sense* definition. In these cases, the law source does not define the term, but instead uses the common meaning of the word, and expands upon that meaning. For example:

| **Protection of Antarctica Act, article 1. sub 2, introduction and sub a** |
| In this law and the stipulations based on it is also understood by Antarctic environment those ecosystems that are dependent on or related to the Antarctic environment. |

They use similar text patterns, but with the inclusion of the word *"ook"* (also) or *"niet"* (not).

Closely related to definitions are the deeming provisions. A deeming provision is a sentence that declares one situation equal to another situation, in a certain context. If one situation is deemed equal to another situation, then the rules that apply to the latter also apply to the first. Deeming provisions introduce some kind of legal fiction. For example:

Income Tax Act 2001, article 2.2, sub 2, introduction and sub a
A Dutchman who is employed by the State of the Netherlands is always deemed to live in the Netherlands if he is posted as a member of a diplomatic, permanent or consular representation of the Kingdom of the Netherlands in foreign countries.

The effect of this statement is that someone is considered to live in the Netherlands, even though he actually lives outside of the Netherlands. In computer models, deeming provisions often appear in the same manner as definitions.

Deeming provisions follow the pattern *"wordt geacht te"* (is deemed to).

4.2 Core Rules and Procedural Rules

The actual content of the legal sources is formed by norms. Normative sentences may confer rights and permissions or impose duties and obligations. Procedural rules are expressed as norms as well, usually they are obligations. Research has been performed on how to distinguish these different types of norms [11]. During this research, it became clear that it was difficult to separate rights and permissions, and that it was likewise difficult to separate duties and obligations. Because of this, these were grouped together. The work yielded a large amount of patterns that were incorporated in the classifier for this project, as they proved useful not only to distinguish between rights and obligations, but also to distinguish these norms from other types of sentences.

An important conclusion was that a majority (almost 71%) of the rights could be identified by the verb *"kunnen"* (may). Another 17% could be identified by the phrase *"is bevoegd"* (is qualified). A host of smaller patterns accounted for the remaining rights.

Sentences denoting duties usually did not follow a pattern; 80% of these sentences was a statement of fact. This means that the text does not state what must happen, but instead simply states that it happens. For example:

Funeral Act, article 46, sub 1
No bodies are interred on a closed cemetery.

In The Netherlands the guidelines for legal drafting recommend the use of the 'statement of fact' instead of a normative formulation [12]. Obviously, such statements of fact have few words in common with similar statements from different domains. Thus, there is no pattern to be found either.

Application Provisions

Application provisions specify situations in which other legislation (usually an article or subsection of an article) does or does not apply. In this way, the application domain of a norm can be extended or restricted (effectively creating an exception to a rule).

Often, an application provision that states that another piece of legislation does apply seems to be included to take away any doubts as to whether it ought to apply or not.

Constitution, article 7, fourth member
The previous members do not apply to making commercial advertisements.

The patterns used by these sentences are "*is (niet) van toepassing*" (does (not) apply).

Penalization

The violation of some norms will carry punishment in the form of a fine or jail. If this is the case, the law will specify the penalties. In Dutch law, this is usually done through sentences like: "*wordt gestraft met*" (is punished with). For example:

Mining Act, article 133, sub 1
Breaking article 43, sub 2, is punished with a monetary fine of the second category.

In general, these sentences are followed by another sentence that denotes whether the punishable fact is a crime or a misdemeanour, such as:

Mining Act, article 133, sub 2
The fact marked as punishable by this article is a misdemeanour.

These sentences always follow this same structure.

Value Assignments and Changes

Any mathematical formulas in a law are given by means of value assignments and changes. A value assignment is a sentence that gives an initial value for a concept value changes are later steps that modify such a value.

An example of such a sentence is:

Income Tax Act 2001, article 3.3, sub 1
Taxable wages are wages reduced with the employee's discount.

These sentences use a range of mathematical operations (to reduce, to increase) and comparisons (at most) which in combination with the verb to be or to amount to can be used to detect them.

4.3 Rule Management

Rule management encompasses a number of sentence types: setting the enactment date, setting the citation title, changing an existing law, and delegating the creating of the new rules.

Enactment Date
This is the first type of sentence that deals with the maintenance of legal texts. All laws will contain a provision for their own enactment date, often relating it to their publication date or deferring that decision to a Royal Decree.

Notaries Act, article 134
This law is enacted on a date to be set by Royal Decree, which may differ for separate parts and articles.

The central pattern used for these sentences is *"treedt in werking op"* (is enacted on), though it is possible to extended this pattern into a number of standardised sentences which are used most of time.

Citation Title
If it is thought necessary, a law will also define a short title which can be used to refer to it.

Notaries Act, article 135
This act may be referred to as: Notaries act.

These sentences follow the standard format *"Deze wet kan worden aangehaald als"* (This act may be referred to as). Together with the enactment date, the citation title is usually present at the end of a legal text.

It is also possible that a law will modify the short title of another law. This is usually done to avoid confusion, when a new law has the same name as he predecessor. In addition, it also happens that the short title may be abbreviated even further. Both types of actions are very rare, and, so far, we have not encountered them often enough to recognise them as a standard type of sentence.

Change Provisions
By means of a change provision, a law can modify some existing legislation. Most laws are change laws, which change some existing legislation rather than introducing a large amount of new rules. In these laws, change provisions make up the bulk of the text.

We distinguish four different types of changes: inserting new text, modifying text, renumbering and deleting or repealing text.

Inserting text follows the pattern *"wordt ingevoegd"* (is inserted) or *"wordt toegevoegd"* (is appended), depending on whether the new text is added somewhere within a section or at the end of the section. An example of such a sentence is:

Act of June 6th, 2002 (Stb. 303), article I, sub IIa
To article 7.36, a new sentence is appended, to read as follows: Article 7.34, sub 5, applies correspondingly.

Modifying text can happen in two ways. If a small amount of text is to be changed, then that text is quoted as well as the new text. The pattern followed is *"wordt vervangen door"* (is replaced by). For example:

> **Act of June 6th, 2002 (Stb. 303), article III, sub V**
> In article 7.12, sub 1, second sentence, «article 7.3b» is replaced by: article 7.3c.

Alternatively, if an entire sentence, section or article is replaced, the modifying provision will simply refer to that element and quote the new text:

> **Act of June 6th, 2002 (Stb. 303), article IV, sub B**
> Article 2.8 will read: …

The pattern for such replacements is *"komt te luiden"* (will read).

A repeal can repeal an entire law, a section of a law or just a bit of a text. It follows the pattern *"vervalt"* (is repealed), as in this example:

> **Act of June 6th, 2002 (Stb. 303), article I, sub QQ**
> Article 17.2 is repealed.

The last change is the renumbering of structural elements. Because renumbering an element requires the modification of all text referring to that element, it is somewhat uncommon for articles to be renumbered. On the other hand, anything below the level of article (subsections and lists) is almost always renumbered to keep a continuous numbering.

Renumbering can either be done explicitly or implicitly. Explicitly renumbering means that the action is identified as a renumbering, using a pattern like *"wordt vernummerd tot"* (is renumbered to) or *"wordt verletterd tot"* (is re-lettered to).

> **Act of June 6th, 2002 (Stb. 303), article IIIc, sub C**
> The articles 17a.1 to 17a.25 are renumbered to the articles 17.20 to 17.54.

Implicit renumbering is done by viewing the number not as an attribute of the structural element, but just as a piece of text, which is subsequently changed.

> **Act of June 6th, 2002 (Stb. 303), article IIIa, sub A**
> The heading of chapter 5a will read: Chapter 5. Accreditation in higher education.

Additionally, renumbering is often not done in a separate provision, but mentioned as a side effect of a provision that inserts new text.

None of the sentences above have a complete reference. They refer to articles, but not to articles in a specific law. Normally, this would mean that they refer to an article in the same text, but this is most often not the case. If many changes are made in the same text, then they are grouped and preceded by a sentence that sets the scope, such as:

> **Act of June 6th, 2002 (Stb. 303), article I, introduction**
> To the Higher Education and Academic Research Act, the following modifications are made:

Such scope declarations follow the pattern *"wordt als volgt gewijzigd"* (is modified as follows) or *"worden de volgende wijzigingen aangebracht"* (the following modifications are made).

Delegation
Delegations confer the power to create additional rules to some legal entity. Most often, this power is conferred onto a minister, for the creation of rules that do not require (immediate) involvement of the parliament. In some cases, the delegation is an order to create rules to arrange for something; in other cases, it merely allows for the creation of rules should the need arise. An example pattern is *"kan regels stellen"* (may create rules).

Agricultural Tenancies Act, article 3, sub 1
By Order in Council, rules are set with regard to the highest allowed rent.

Related to the delegations are the publishing provisions, which order a minister to publish certain rules created by a (lower) entity.

5 Experimental Classifier

We built a classifier (in Java) that takes well structured legal sources as input and tries to classify their sentences according to their type based on typical patterns associated with these types. An important limitation to this approach is formed by the statements of fact. As stated in section 4, these do not follow any recognisable patterns. Because of this, we hope to identify this important group of statements "by default": if we can identify patterns for everything else, we may assume that anything not classified by these patterns is one of these statements of fact that actually reflect norms.

The classifier assumes that the input is structured using MetaLex XML. In MetaLex, sentences and lists are marked, as well as the separate list items within each list. This enables the classifier to treat each sentence separately. This means that a necessary pre-condition for the classifier is that the structure of the documents has already been marked. For legacy texts an automatic structure recogniser would therefore be desirable. For more recent texts, this is usually not a problem. These documents often have already been tagged, or have even been produced in XML format using modern XML based legislative editors like MetaVex or XMLeges[3].

The classifier is a simple pattern matcher. We used 88 patterns from about twenty Dutch laws. Most patterns consist of only a verb phrase, like *"mogen"* (may) for a right/permission or *"wordt aangehaald als"* (is referred to as) for the defining of a citation title. Sometimes, additional keywords have been added, as in *"kan regels stellen"* (may create rules).

The patterns are stored in a format for the Java pattern matcher (`java.util.regex`). The patterns mentioned above become:

---

[3] In addition to automatically marking the structure, such editors could even tag sentences during construction as being of a specific type, for instance because the legislative drafter chooses a particular template in MetaVex [13].

```
\s+(mag|mogen)\b
\s+wordt\s+aangehaald\s+als(:)?\s+
\s+kan\s+regels\+stellen\s+
```

In this format, \s+ denotes one or more whitespace characters, \b denotes a word boundary, and (:)? is an optional colon. The first pattern allows for either the singular or the plural form of the verb.

The classifier will attempt to match a sentence to each available pattern. If the sentence matches several patterns, the classifier will prefer the longest of the matches. (This does not happen often; however, some of the patterns overlap, such as *"kan"* for a right and *"kan regels stellen"* for a delegation).

A specific strategy needed to be chosen to tackle embedded lists, such as.

Tobacco Act, article 1

In this law, and in the stipulations based on it, is understood by:

a.   tobacco products: ... ;

b.   Our Minister: ...;

c.   appendix: ...;

...

Our initial assumption was that we could base the classification on the first part of the sentence, and that the individual list items were not needed for the classifications. This assumption did not hold, as we encountered list in which the verbs, and thus the patterns, did not occur in that first sentence part [14]. An example of such a list is:

Bill 20 585 nr.2, article 5

Our Ministers:

a. appoint, suspend and discharge the chairman and other members, after hearing the council involved;

b. appoint, suspend and discharge the advising members.

As the first sentence part does not contain any patterns, these lists were always classified as a statement of fact (the default) which, in the case of this example, would have been correct.

To come to a more correct classification, we came to a different approach: the classifier first classifies the introduction. If it does not find an explicit pattern, it then continues to classify the different list items.

If the first sentence does contain an explicit pattern, the list would be classified as entirely of the matching type. If the first sentence does not contain an explicit pattern, but (some of) the list items did, the list items would be classified independently. If neither contained explicit patterns, the list as a whole would be classified as a statement of fact.

Another group of somewhat difficult sentences are those that replace or insert text. Those sentences quote old and/or new text fragments, which also may contain a pattern. Thus, the sentence as a whole can contain multiple patterns and can easily be misclassified.

Because we assume that the input is tagged in MetaLex, this does not pose an actual problem, though. In MetaLex, such texts are marked using so-called 'quote' elements. When a sentence is parsed that contains such elements, the contents of those quotes are not used in classifying the sentence. If the quoted text contains *complete* sentences or lists however, the classifier will attempt to classify those as well as the containing sentence.

6 Results

We tested the classifier on eighteen different Dutch regulations, different from the twenty we used to come up with the patterns. Four of these eighteen regulations were completely new laws; the others changed already existing laws (as is the more common situation). With the exception of a single Royal Decree, these where all bills pending at Parliament. In this they differed from the set used to derive the patterns, which where all acts that had already been passed. The length of the laws varied from very short (three sentences) to quite long (166 sentences on 23 pages A4); most were quite recent (patterns in the past have been different).

All laws are listed in Table 1 below. To check whether clauses were classified correctly, all sentences and lists in all laws were also classified manually.

Table 1. Results per source

Source	Sentence			List				Type
	Total	Correct	%	Total	Correct	Partial	%	
Royal Decree Stb. 1945, F 214 (as modified per 01/01/2002)	26	23	97%	4	4	0	75%	New
Bill 20 585 nr. 2	31	30	97%	4	3	1	75%	New
Bill 22 139 nr. 2	22	20	91%	2	2		100%	New
Bill 27 570 nr. 4	21	16	76%					Change
Bill 27 611 nr. 2	11	11	100%	1	1		100%	Change
Bill 30 411 nr. 2	141	128	91%	25	20	3	80%	New
Bill 30 435 nr. 2	40	39	98%	4	3	1	75%	Change
Bill 30 583 nr. A	27	27	100%					Change
Bill 31 531 nr. 2	3	3	100%					Change
Bill 31 537 nr. 2	29	29	100%	2	2	0	100%	Change
Bill 31 540 nr. 2	7	7	100%					Change
Bill 31 541 nr. 2	8	8	100%					Change
Bill 31 713 nr. 2	7	6	86%	2	2	0	100%	Change
Bill 31 722 nr. 2	31	22	71%	6	5	0	83%	Change
Bill 31 726 nr. 2	78	67	86%	2	1	1	50%	Change
Bill 31 832 nr. 2	7	7	100%	3	3		100%	Change
Bill 31 833 nr. 2	4	4	100%					Change
Bill 31 835 nr. 2	99	90	91%	7	4	3	57%	Change
Total	**592**	**537**	91%	**62**	**50**	**9**	81%	

Table 1 shows the total number of sentences and lists in the sources we used for testing, as well as the number of sentences and lists that were classified correctly. The column *type* indicates whether a law was completely new, or whether it was a change law, mainly aimed at modifying an existing piece of legislation. Change laws are far more common than completely new legislation. They tend to contain less definitions and norms than new legislation.

The classifier performs well, classifying 91% of all sentences and 81% of all lists correctly[4]. We expect these results to generalize over all Dutch laws. For most natural language domains, a pattern based systems is thought to have too little generalisation capacity. Although languages do have underlying rules, people will often stretch and bend these to their need. As a result, a system based upon patterns is often too rigid to deal with all the information that can occur [16], and a statistical method is usually recommended [16,17]. From our results, it becomes clear that most laws use only a limited set of patterns. If a pattern is missing, the accuracy of the classifier could drop fast. However, the amount of variation in legal texts is restricted, as legal drafters will seldom use a completely new style, instead using the style of older laws or the official guidelines. This is also confirmed in our tests: the majority of all sentences is classified by a small number of patterns.

On lists, the classifier does perform a lot worse than on sentences. Many lists that should be classified as a statement of fact, and that should be classified based on the first sentence part, were misclassified in this new approach. This was due to the fact that the list items contained one of the other patterns, usually as a subordinate clause. (Subordinate clauses lead to more problems, which we will discuss below).

In most cases, the first sentence part is supposed to form a correct sentence with each of the separate list items. In order to classify a list, we could derive each of those sentences and classify them. This would solve some of the problems, when the pattern was split over the introduction and the individual list items.

Table 2 presents the results for the different types of sentences (as the performance on lists depends very much on the approach chosen on how to handle them, and not only on the patterns, we will keep them out of the discussions on the performance of the patterns). The column *"In corpus"* shows the number of sentences present in the test set for each type, both as an absolute number and as a percentage. The column *"Missed"* shows how many of these sentences were not correctly identified. For example, the test set contained 40 application provisions, but one was incorrectly classified (meaning that 39 were correctly classified). The column *"False"* presents the amount of sentences that were incorrectly classified as a particular type, e.g. in the same row as before eight sentences were incorrectly classified as an application provision. Each false positive corresponds to a 'missed' somewhere else.

The greatest part of all the sentences is formed by the norms. 43% of all sentences belong to one of the norm categories. The next biggest category consists of the change provisions, with 41%. These changes can be further broken down as follows:

---

[4] Which is actually worse than the method we tested in [19], where we classified all lists based on their introduction sentence.

Table 2. Results per sentence type

Type	In corpus		Missed	False
Definition	2%	12	1	0
Norm - Right/Permission	11%	64	4	13
Norm - Obligation/Duty	5%	29	0	1
Delegation	3%	19	6	0
Publication Provision	1%	4	0	0
Application Provision	7%	40	1	8
Enactment Date	3%	17	1	0
Citation Title	1%	3	0	0
Value Assignment	0%	1	0	0
Penalisation	0%	0	0	2
Change	41%	241	16	8
Mixed Type	1%	3	3	0
Norm - Statement of Fact (default)	27%	159	23	23
Total		592	55	55

Table 3. Results for change sentence types

Type	In corpus		Missed	False
Scope	9%	54	0	0
Insertion	7%	44	1	0
Replacement	19%	111	4	0
Repeal	4%	23	7	8
Renumbering	2%	9	4	0
Total		241	16	8

Some sentences were a concatenation of two sentences. For example, one sentence contained two changes: a renumbering and a repeal. These sentences are listed in Table 2 as 'Mixed type'. About half of the misses were caused by patterns that were unknown to the classifier. These sentences were incorrectly classified as the default (statement of fact), and sometimes as a norm of the type right/permission.

Two notable patterns were missing: a renumbering pattern dealing with re-lettering rather than renumbering, and a new pattern for delegations. Both of these should probably be added to the classifier. The other missing patterns occurred only once, and unless we encounter them much more often in the future, these are errors that are unlikely to be repaired in future versions of the classifier.

Those misclassifications that were not caused by missing patterns were instead caused by patterns that were somehow too broad. For example, most false positives of the "repealed" type sentences were provisions concerning the repeal of fines instead of articles. This will require more sophisticated patterns or dedicated 'anti-patterns' (i.e. not applicable when it contains the word 'fine').

Both false penalisations were in fact a right; the pattern that triggered this classification was part of a qualification of a legal body that was given certain rights. Such a qualification is given in a subordinate sentence. This means that the classifier will find two (or even more) patterns: one in the subordinate sentence, and one in the main sentence. As it does not have the option to distinguish between the two, it will pick the longest match (which will not always be the correct one).

If the main sentence does not contain any pattern (because it is a statement of fact), the classifier will only find the pattern in the subordinate sentence, and will automatically arrive at the wrong conclusion. This is the cause of almost all false rights and false application statements. It would be preferable if the classifier could ignore the subordinate sentences completely. This would require that the sentences be split into main and subordinate sentences before classification, which would mean creating a far heavier application than the current classifier. However, as the classification is the first step in the larger process to automatically generate models, splitting main and subordinate sentences will be of use (most likely even necessary) in later steps of the process.

We only encountered one value assignment in the texts classified during this experiment. These seem to be specific to certain domains (i.e. taxes), and perhaps they are usually deferred to lower order regulations. The test set did not contain any deeming provisions, nor any penalisation provisions.

Of the 88 patterns we identified in our training set, only 44 were actually present in our test set. A possible explanation for this is that our training set was more spread out in time, while our test set consists mainly of rather recent laws (not counting the Royal Decree from 1945, the laws in the test set are all from the last 20 years, whereas the training set contains several laws from before 1960, going back as far as 1875). However, these test results may also be an indication that there are too many patterns in our classifier, and that some ought to be removed.

Table 4 shows the distribution of the patterns that were actually encountered in the test set. The numbers suggest that there are a couple of main patterns that account for a majority of the sentences identified. For example, of the 60 correctly identified rights, 55 used the pattern "may" and four used the pattern "is qualified". This corresponds to Franssen's [11] conclusion that the majority of rights could be identified with those two patterns. This distribution again suggests that some of the other patterns may be superfluous.

Table 4 also shows that most of the false positives are caused by a small set of patterns: one pattern for repeal, one for rights and one for application provisions. A possible solution to the false positives may be to narrow these patterns down. However, these three patterns are also responsible for a large number of correct identifications, and narrowing them down may reduce the success rate.

If we study the number of patterns encountered in each separate law, shown in Table 5, we see that most laws do not use a lot of different patterns. One or two per type seems most common, with sometimes three or four patterns for the bigger categories. This may be explained by the fact that a limited number of legal drafters work

Table 4. Distribution of patterns used

Type	Patterns Known	Patterns Used	Results per pattern Correct	False
Definition	14	5	6	0
			2	0
			1	0
			1	0
			1	0
Norm – Right/Permission	17	3	55	13
			4	0
			1	0
Norm - Obligation/Duty	15	8	6	1
			5	0
			3	0
			2	0
			2	0
			1	0
			1	0
			1	0
Delegation	7	5	5	0
			4	0
			2	0
			1	0
			1	0
Publication Provision	1	1	4	0
Application Provision	5	5	36	5
			2	0
			1	0
			0	1
			0	2
Enactment Date	1	1	16	0
Citation Title	2	2	2	0
			1	0
Value Assignment	8	1	1	0
Penalisation	3	1	0	2
Change – Scope	2	2	49	0
			5	0
Change – Insertion	4	4	22	0
			18	0
			2	0
			1	0
Change - Replacement	3	3	66	0
			40	0
			1	0
Change – Repeal	2	1	16	8
Change - Renumbering	3	2	4	0
			1	0
Total	**87**	**44**	**393**	**32**

on a specific law (on the really small ones perhaps only one) and have their specific sets of regular expressions they use.

This suggests that if a pattern that is used in a law is missing in the classifier, there will be a huge drop in the accuracy of the classifier for that type, as most patterns account for a fairly large portion of the sentences. During this test, this has only occurred in small categories, so this had only a limited impact (i.e. in Bill 31 835, all five repealing sentences have been missed due to a missing pattern).

Table 5. Experimental results for all bills

	Number of sentences	Definition	Norm - Right/Permission	Norm - Obligation/Duty	Delegation	Publication Provision	Application Provision	Enactment Date	Citation Title	Value Assignment	Change - Scope	Change - Insertion	Change - Replacement	Change - Repeal	Change - Renumbering
Number of patterns		14	17	15	7	1	5	1	2	8	2	4	3	2	3
Royal Decree Stb. 1945, F 214	26		1	3	3		2	1						1	
Bill 20 585 nr. 2	31		2	1			1	1	1						
Bill 22 139 nr. 2	22		2	1			1	1			1		1		
Bill 27 570 nr. 4	21		1					1			1		2	1	1
Bill 27 611 nr. 2	11		1	2				1			1	1	1		
Bill 30 411 nr. 2	141	4	2	3	2	1	2	1	1	1	1	2	2	1	
Bill 30 435 nr. 2	40		1				2	1	1		1	2	2	1	1
Bill 30 583 nr. A	27			1				1			1	2	1		
Bill 31 531 nr. 2	3						1	1					1		
Bill 31 537 nr. 2	29			2			1	1			2	2		1	
Bill 31 540 nr. 2	7	1						1			1	1		1	1
Bill 31 541 nr. 2	8						1	1			1	1	1		
Bill 31 713 nr. 2	7						1	1			1	1	1	1	
Bill 31 722 nr. 2	31		1		1		1	1			1	1	2	1	
Bill 31 726 nr. 2	78		1	1	2		1				1	2	2	1	1
Bill 31 832 nr. 2	7	1						1					1		
Bill 31 833 nr. 2	4						1	1				1			
Bill 31 835 nr. 2	99		1		1		1	1			1		2	1	

7 Conclusions and Discussion

In this chapter, we have discussed the detection and classification of norms in legal texts using normative sentences combined with sentences that form definitions, deeming provisions, exceptions and application provisions. We distinguished between different layers of norms in a more detailed extension of Hart's primary and secondary norms. In a lot of existing models and applications, deeming provisions, exceptions and application provisions are not retained, but simplified to if-then-else statements. That is why, at the start of this chapter, we proposed to make an intermediate model of the law that retains isomorphism with the original text and would be easier to maintain.

A first step to the creation of such models is classifying the provisions that occur in a law. In Section 4, we have presented a possible classification. The classification presented there seems to be adequate for Dutch laws. All provisions encountered while gathering information are covered, as well as all provisions encountered in our test data. The classification is based on Dutch law texts. In order to extend it to other jurisdictions, some modification may be needed. However, a comparison with Tiscornia and Turchi [15] suggests that the classification for Dutch and Italian law is rather similar.

Although there are multiple language constructs for each sentence type, these are limited, making them easy to detect. As an experiment, we have set up a classifier that attempts to classify sentences based on these patterns. That classifier works well. Within the laws used for our experiment, 91% of all phrases were classified correctly, with hardly any false positives. Almost 43% of all phrases were classified as some type of norm, a further 41% as clauses changing an existing law. These results are similar to those reached in [9], where a machine learning approach is used to classify provisions in Italian law. This suggests that both methods are capable of reaching comparable results, despite the fact that patterns are often seen as less suitable for dealing with natural language. We suspect that the issues that arise with a pattern-based approach, like the errors generated by auxiliary sentences, will also hamper a machine learning approach.

We expect these results to generalize over all Dutch laws, though we need more patterns for specific domains (see below). For other languages and jurisdictions, we expect that the same categories, with different language patterns, will form a good starting point.

Of course, there is room for improvement as well. A major cause of misclassifications is the occurrence of patterns in subordinate sentences. Detecting these sentences beforehand so they may be ignored during the classification would lead to a serious increase in performance. Another improvement can be achieved in handling the lists. Both straightforward methods we have tested did not perform as well as we would desire (though the recognition of subordinate sentences would help here as well).

Finally, additional patterns may be added to improve results. We know that specific domains (such as tax law) will be needing specific patterns. For example, in our earlier research on the Income Tax Act 2001, the patterns used for definitions and type extensions were: *x is y*, or: *x are y* instead of *by x is understood y* 10. So far, our results suggest that the Income Tax Act was unique in its use of these patterns, and we have not included them in the current version of the classifier (meaning that this classifier will not perform well on the Tax Income Act for definitions). Also, some of the sentences that were not classified in the test used language constructs that may also be new patterns, which should be added if they are encountered more often.

The next step should be to create actual models from a sentence, using its classification as a basis.

Acknowledgements

These experiments and their results were presented in two earlier conference papers: [14] and [22]. An earlier version of the categorization of norms was presented in [19]. We would like to thank our student Gijs Kruitbosch for his work on the classifier.

References

1. de Maat, E., Winkels, R., van Engers, T.: Automated Detection of Reference Structures in Law. In: van Engers, T. (ed.) Legal Knowledge and Information Systems - JURIX 2006: The Nineteenth Annual Conference, pp. 41–50. IOS Press, Amsterdam (2006)
2. van Engers, T.M., Kordelaar, P.J.M., Den Hartog, J., Glassée, E.: POWER: Programme for an ontology based working environment for modelling and use of regulations and legislation. In: Tjoa, A.M., Wagner, R.R., Al-Zobaidie, A. (eds.) Proceedings of the 11th workshop on Databases and Expert Systems Applications (IEEE), Greenwich, London, pp. 327–334 (2000)
3. Hart, H.: The Concept of Law. Clarendon Press, Oxford (1961)
4. Ross, A.: Directives and norms. Routledge and Kegan Paul Ltd., England (1968)
5. van Engers, T.M., Winkels, R., Boer, A., de Maat, E.: Internet, portal to justice? In: Gordon, T. (ed.) Legal Knowledge and Information Systems. Jurix 2004: The Seventeenth Annual Conference. Frontiers in Artificial Intelligence and Applications, pp. 131–140. IOS Press, Amsterdam (2004)
6. Winkels, R., Boer, A., de Maat, E., van Engers, T., Breebaart, M., Melger, H.: Constructing a semantic network for legal content. In: Gardner, A. (ed.) Proceedings of the Tenth International Conference on Artificial Intelligence and Law (ICAIL 2005), pp. 125–140. ACM Press, New York (2005)
7. Bench-Capon, T.J.M., Coenen, F.P.: Isomorphism and Legal Knowledge Based Systems. Artificial Intelligence and Law 1(1), 65–86 (1992)
8. Hoekstra, R., Breuker, J., Di Bello, M., Boer, A.: The LKIF Core ontology of basic legal concepts. In: Casanovas, P., Biasotti, M.A., Francesconi, E., Sagri, M.T. (eds.) Proceedings of the 2nd Workshop on Legal Ontologies and Artificial Intelligence Techniques (LOAIT 2007). CEUR-WS.org, pp. 43–63 (2007)
9. Biagioli, C., Francesconi, E., Passerini, A., Montemagni, S., Soria, C.: Automatic semantics extraction in law documents. In: Gardner, A. (ed.) Proceedings of the Tenth International Conference on Artificial Intelligence and Law (ICAIL 2005), pp. 133–140. ACM Press, New York (2005)
10. de Maat, E.: Natural Legal Modelling. Master's thesis, University of Twente, Enschede (2003)
11. Franssen, M.: Automated Detection of Norm Sentences in Laws. In: Twente Student Conference on IT (2007)
12. Aanwijzigingen voor de regelgeving. Circulaire van de Minister-President. Original in Staatscourant 1992, 230 (1992); Last modification in Staatscourant 2005, 58 (2005)

13. van de Ven, S., Hoekstra, R., Winkels, R., de Maat, E., Kollár, A.: MetaVex: Regulation Drafting Meets the Semantic Web. In: Casanovas, P., Sartor, G., Casellas, N., Rubino, R. (eds.) Computable Models of the Law. LNCS (LNAI), vol. 4884, pp. 42–55. Springer, Heidelberg (2008)

14. de Maat, E., Winkels, R.: Automatic Classification of Sentences in Dutch Laws. In: Francesconi, E., Sartor, G., Tiscornia, D. (eds.) Legal Knowledge and Information System - JURIX 2008: The 21st Annual Conference Annual Conference, pp. 207–216. IOS Press, Amsterdam (2008)

15. Tiscornia, D., Turchi, F.: Formalization of legislative documents based on a functional model. In: Proceedings of the 6th international conference on Artificial intelligence and law, pp. 63–71. ACM Press, New York (1997)

16. Nanning, C.D., Schütze, H.: Foundations of Statistical Natural Language Processing. MIT Press, Cambridge (1999)

17. Moens, M.-F.: Innovative techniques for legal text retrieval. Artificial Intelligence and Law 9(1), 29–57 (2001)

18. de Maat, E., Winkels, R., van Engers, T.: Making Sense of Legal Texts. In: Grewendorf, G., Rathert, M. (eds.) Formal Linguistics and Law, Mouton, De Gruyter (2010)

19. de Maat, E., Winkels, R.: Categorisation of Norms. In: Lodder, A., Mommers, L. (eds.) Legal Knowledge and Information Systems - JURIX 2007: The Twentieth Annual Conference, pp. 79–88. IOS Press, Amsterdam (2007)

20. Gonçalves, T., Quaresma, P.: Is linguistic information relevant for the classification of legal texts? In: Gardner, A. (ed.) Proceedings of the Tenth International Conference on Artificial Intelligence and Law (ICAIL 2005), pp. 168–176. ACM Press, New York (2005)

21. McCarty, T.: Deep Semantic interpretations of legal texts. In: Winkels, R. (ed.) Proceedings of ICAIL 2007, pp. 217–224. ACM Press, New York (2007)

22. de Maat, E., Winkels, R.: A Next Step towards Automated Modelling of Sources of Law. In: Hafner, C. (ed.) Proceedings of ICAIL 2009, pp. 31–39. ACM Press, New York (2009)

Efficient Multilabel Classification Algorithms
for Large-Scale Problems in the Legal Domain

Eneldo Loza Mencía and Johannes Fürnkranz

Knowledge Engineering Group
Technische Universität Darmstadt
{eneldo,juffi}@ke.tu-darmstadt.de

Abstract. In this paper we apply multilabel classification algorithms to
the EUR-Lex database of legal documents of the European Union. For
this document collection, we studied three different multilabel classification problems, the largest being the categorization into the EUROVOC
concept hierarchy with almost 4000 classes. We evaluated three algorithms: (i) the binary relevance approach which independently trains one
classifier per label; (ii) the multiclass multilabel perceptron algorithm,
which respects dependencies between the base classifiers; and (iii) the
multilabel pairwise perceptron algorithm, which trains one classifier for
each pair of labels. All algorithms use the simple but very efficient perceptron algorithm as the underlying classifier, which makes them very
suitable for large-scale multilabel classification problems. The main challenge we had to face was that the almost 8,000,000 perceptrons that had
to be trained in the pairwise setting could no longer be stored in memory.
We solve this problem by resorting to the dual representation of the perceptron, which makes the pairwise approach feasible for problems of this
size. The results on the EUR-Lex database confirm the good predictive
performance of the pairwise approach and demonstrates the feasibility
of this approach for large-scale tasks.

Keywords: Text Classification, Multilabel Classification, Legal
Documents, EUR-Lex Database, Learning by Pairwise Comparison.

1 Introduction

The EUR-Lex text collection is a collection of documents about European Union
law. It contains many different types of documents, including treaties, legislation, case-law and legislative proposals, which are indexed according to several
orthogonal categorization schemes to allow for multiple search facilities. The
most important categorization is provided by the EUROVOC descriptors, which
is a topic hierarchy with almost 4000 categories regarding different aspects of
European law.

This document collection provides an excellent opportunity to study text classification techniques for several reasons:

- it contains multiple classifications of the same documents, making it possible to analyze the effects of different classification properties using the

E. Francesconi et al. (Eds.): Semantic Processing of Legal Texts, LNAI 6036, pp. 192–215, 2010.
© Springer-Verlag Berlin Heidelberg 2010

same underlying reference data without resorting to artificial or manipulated classifications,

- the overwhelming number of produced documents make the legal domain a very attractive field for employing supportive automated solutions and therefore a machine learning scenario in step with actual practice,
- the documents are available in several European languages and are hence very interesting e.g. for the wide field of multi- and cross-lingual text classification,
- and, finally, the data is freely accessible (at `http://eur-lex.europa.eu/`)

In this paper, we make a first step towards analyzing this database by applying multilabel classification techniques on three of its categorization schemes. The database is a very challenging multilabel scenario due to the high number of possible labels (up to 4000), which, for example, exceeds the number of labels in the REUTERS databases by one order of magnitude. The EUR-Lex dataset is now publicly available under `http://www.ke.tu-darmstadt.de/resources/eurlex/`.

We evaluated three methods on this task:

- the conventional binary relevance approach (BR), which trains one binary classifier per label
- the *multilabel multiclass perceptron* (MMP), which also trains one classifier per label but does not treat them independently, instead it tries to minimize a ranking loss function of the entire ensemble [1]
- the *multilabel pairwise perceptron* (MLPP), which trains one classifier for each pair of classes [2]

Previous work on using these algorithms for text categorization [2] has shown that the MLPP algorithm outperforms the other two algorithms, while being slightly more expensive in training (by a factor that corresponds to the average number of labels for each example). However, another key disadvantage of the MLPP algorithm is its need for storing one classifier for each pair of classes. For the EUROVOC categorization, this results in almost 8,000,000 perceptrons, which would make it impossible to solve this task in main memory.

To solve this problem, we introduce and analyze a novel variant that addresses this problem by representing the perceptron in its dual form, i.e. the perceptrons are formulated as a combination of the documents that were used during training instead of explicitly as a linear hyperplane. This reduces the dependence on the number of classes and therefore allows the *Dual MLPP* algorithm to handle the tasks in the EUR-Lex database.

Originally, the MLPP accepts multilabel information but only outputs a ranking over all possible labels, following [1] and their MMP algorithm. In order to find a delimiter between relevant and irrelevant labels within a provided ranking of the labels, we have recently introduced the idea of using an artificial label that encodes the boundary between relevant and irrelevant labels for each example [3], which has also been successfully applied to the Reuters-RCV1 text categorization task [4], a large collection of news texts. This approach was adapted to work with the dual variant and we present first results in this paper. However, we will focus our analysis on the produced ranking. There are three reasons for this: (i) the MMP, to which we directly compare, and the pairwise method naturally provide such a ranking, (ii) the ranking allows to evaluate the performance differences on a finer scale, (iii) our key motivation is to study the scalability of these approaches which is determined by the rankings, and (iv) although several different thresholding techniques exist that can be applied to the rankings produced by both MMP and MLPP (a good overview is provided in [5]), it was not the intention of this work to provide a comparison between them.

The outline of the paper is as follows: We start with a presentation of the EUR-Lex respository and the datasets that we derived from it (Section 2). Section 3 briefly recapitulates the algorithms that we study, followed by the presentation of the dual version of the MLPP classifier (Section 4). In Section 5, we compare the computational complexity of all approaches, and present the experimental results in Section 6.

2   The EUR-Lex Repository

The EUR-Lex/CELEX (Communitatis Europeae LEX) Site[1] provides a freely accessible repository for European Union law texts. The documents include the official Journal of the European Union, treaties, international agreements, legislation in force, legislation in preparation, case-law and parliamentary questions. They are available in most of the languages of the EU, and in the HTML and PDF format. We retrieved the HTML versions with bibliographic notes recursively from all (non empty) documents in the English version of the *Directory of Community legislation in force*[2], in total 19,348 documents. Only documents related to secondary law (in contrast to primary law, the constitutional treaties of the European Union) and international agreements are included in this repository. The legal form of the included acts are mostly *decisions* (8,917 documents), *regulations* (5,706), *directives* (1,898) and *agreements* (1,597). This version of the dataset differs slightly from that presented in previous works [6, 7], which still contained 19,596 documents. Some empty documents that were missed in the previous version and all corrigendums (they contained the same standard text except for one document since they were concerned with translations of the law into other languages than English) have been removed. The updated version can be found under http://www.ke.tu-darmstadt.de/resources/eurlex/.

The bibliographic notes of the documents contain information such as dates of effect and validity, authors, relationships to other documents and classifications. The classifications include the assignment to several EUROVOC descriptors, directory codes and subject matters, hence all classifications are multilabel ones. EUROVOC is a multilingual thesaurus providing a controlled vocabulary for European Institutions[3]. Documents in the documentation systems of the EU are indexed using this thesaurus. The directory codes are classes of the official

---

[1] http://eur-lex.europa.eu
[2] http://eur-lex.europa.eu/en/legis/index.htm
[3] http://europa.eu/eurovoc/

Title and reference

Council Directive 91/250/EEC of 14 May 1991 on the legal protection of computer programs

Classifications

EUROVOC descriptor
- *data-processing law, computer piracy, copyright, software, approximation of laws*

Directory code
- 17.20.00.00 *Law relating to undertakings* / Intellectual property law

Subject matter
- *Internal market, Industrial and commercial property*

Text

COUNCIL DIRECTIVE of 14 May 1991 on the legal protection of computer programs (91/250/EEC)

THE COUNCIL OF THE EUROPEAN COMMUNITIES,

Having regard to the Treaty establishing the European Economic Community and in particular Article 100a thereof,

Having regard to the proposal from the Commission (1),

In cooperation with the European Parliament (2),

...

Fig. 1. Excerpt of a EUR-Lex sample document with the CELEX ID 31991L0250. The original document contains more meta-information. We trained our classifiers to predict the EUROVOC descriptors, the directory code and the subject matters based on the text of the document.

classification hierarchy of the *Directory of Community legislation in force*. It contains 20 chapter headings with up to four sub-division levels.

A large number of 3,956 different EUROVOC descriptors were identified in the retrieved documents. Each document is associated to 5.31 descriptors on average. In contrast there are only 201 different subject matters appearing in the dataset, with a mean of 2.21 labels per document, and 410 different directory codes, with a label set size of on average 1.29. Note that for the directory codes we used only the assignment to the leaf category as the parent nodes can be deduced from the leaf node assignment. For the document in Figure 1 this would mean a set of labels of {17.20} instead of {17, 17.20}. An overview of the properties of the different views on the dataset are given in Table 1.

Figure 1 shows an excerpt of a sample document with all information that has not been used removed. The full document can be viewed at http://eur-lex.europa.eu/LexUriServ/LexUriServ.do?uri= CELEX:31991L0250:EN:NOT. We extracted the text body from the HTML documents, excluding HTML tags, bibliographic notes or other additional information that could distort the results. The text was tokenized into lower case, stop words

Table 1. Statistics of datasets. The attribute number in parenthesis denotes the actual used number of features, i.e. for *scene* and *yeast* the number of features after adding the pairwise products and for the text collections the amount after feature selection. *Label density* indicates the average number of labels per instance d relative to the total number of classes n, and *distinct* counts the distinct label-sets found in the dataset $|\{P_i \mid i = 0 \ldots m\}|$.

dataset name	#classes n	avg. label-set size d	density $\frac{d}{n}$	distinct
EUR-Lex *subject matter*	201	2.213	1.101 %	2540
EUR-Lex *directory code*	410	1.292	0.315 %	1615
EUR-Lex *EUROVOC*	3956	5.310	0.134 %	16467

were excluded, and the Porter stemmer algorithm was applied. In order to perform cross validation, the instances were randomly distributed into ten folds. The tokens were projected for each fold into the vector space model using the common TF-IDF term weighting.In order to reduce the memory requirements, of the approx. 200,000 resulting features we selected the first 5,000 ordered by their document frequency. This feature selection method is very simple and efficient and independent from class assignments, although its performance is comparable to more sophisticated methods using chi-square or information gain computation [8]. In order to ensure that no information from the test set enters the training phase, the TF-IDF transformation and the feature selection were conducted only on the training sets of the ten cross-validation splits.

The EUROVOC thesaurus has already been presented as set of classes for a multilabel classification task in [9]. The authors use several refined text and linguistig processing techniques and statistical computations in order to return a list of associated lemmas from the EUROVOC thesaurus for documents of the EU. However, their results are not comparable since a different resource was used for the documents, resulting also in a different number of EUROVOC descriptors used, namely around 2900.

3   Preliminaries

We represent an instance or object as a vector $\bar{\mathrm{x}} = (x_1, \ldots, x_N)$ in a feature space $\mathcal{X} \subseteq \mathbb{R}^N$. Each instance $\bar{\mathrm{x}}_i$ is assigned to a set of relevant labels P_i, a subset of the n possible classes $\mathcal{L} = \{\lambda_1, \ldots, \lambda_n\}$. For multilabel problems, the cardinality $|P_i|$ of the label sets is not restricted, whereas for binary problems $|P_i| = 1$. For the sake of simplicity we use the following notation for the binary case: we define $\mathcal{L} = \{1, -1\}$ as the set of classes so that each object $\bar{\mathrm{x}}_i$ is assigned to a $\lambda_i \in \{1, -1\}$, $P_i = \{\lambda_i\}$.

3.1   Ranking Loss Functions

In order to evaluate the predicted ranking we use different *ranking losses*. The losses are computed comparing the ranking with the true set of relevant classes, each of them focusing on different aspects. For a given instance $\bar{x}$, a relevant label set P, a negative label set $N = \mathcal{L} \backslash P$ and a given predicted ranking $r : \mathcal{L} \to \{1 \ldots n\}$, with $r(\lambda)$ returning the position of class λ in the ranking, the different loss functions are computed as follows:

- The is-error loss (IsErr) determines whether $r(\lambda) < r(\lambda')$ for all relevant classes $\lambda \in P$ and all irrelevant classes $\lambda' \in \overline{P}$. It returns 0 for a completely correct, *perfect ranking*, and 1 for an incorrect ranking, irrespective of 'how wrong' the ranking is.
- The one-error loss (OneErr) is 1 if the top class in the ranking is not a relevant class, otherwise 0 if the top class is relevant, independently of the positions of the remaining relevant classes.
- The ranking loss (RankLoss) returns the number of pairs of labels which are not correctly ordered normalized by the total number of possible pairs. As IsErr, it is 0 for a perfect ranking, but it additionally differentiates between different degrees of errors.

$$E \stackrel{\text{def}}{=} \{(\lambda, \lambda') \mid r(\lambda) > r(\lambda')\} \subseteq P \times N \qquad \delta_{\text{RankLoss}} \stackrel{\text{def}}{=} \frac{|E|}{|P||N|} \qquad (1)$$

- The margin (Margin) loss returns the number of positions between the worst ranked positive and the best ranked negative classes. This is directly related to the number of wrongly ranked classes, i.e. the positive classes that are ordered below a negative class, or vice versa. We denote this set by F.

$$F \stackrel{\text{def}}{=} \{\lambda \in P \mid r(\lambda) > r(\lambda'), \lambda' \in N\} \cup \{\lambda' \in N \mid r(\lambda) > r(\lambda'), \lambda \in P\} \qquad (2)$$

$$\delta_{\text{Margin}} \stackrel{\text{def}}{=} \max(0, \max\{r(\lambda) \mid \lambda \in P\} - \min\{r(\lambda') \mid \lambda' \notin P\}) \qquad (3)$$

- Average Precision (AvgP) is commonly used in Information Retrieval and computes for each relevant label the percentage of relevant labels among all labels that are ranked before it, and averages these percentages over all relevant labels. In order to bring this loss in line with the others so that an optimal ranking is 0, we revert to the following measure.

$$\delta_{\text{AvgP}} \stackrel{\text{def}}{=} 1 - \frac{1}{P} \sum_{\lambda \in P} \frac{|\{\lambda^* \in P \mid r(\lambda^*) \le r(\lambda)\}|}{r(\lambda)} \qquad (4)$$

3.2 Multilabel Evaluation Measures

There is no generally accepted procedure for evaluating multilabel classifications. Our approach is to consider a multilabel classification problem as a meta-classification problem where the task is to separate the set of possible labels into relevant labels and irrelevant labels. Let $\hat{P}_i$ denote the set of labels predicted by the multilabel classifier and $\hat{N}_i = \mathcal{L} \setminus \hat{P}_i$ the set of labels that are not predicted by the classifier for an instance $\bar{x}_i$. Thus, we can, for each individual instance $\bar{x}_i$, compute a two-by-two confusion matrix C_i of relevant/irrelevant vs. predicted/not predicted labels:

C_i	predicted	not predicted	
relevant	$\lvert P_i \cap \hat{P}_i \rvert$	$\lvert P_i \cap \hat{N}_i \rvert$	$\lvert P_i \rvert$
irrelevant	$\lvert N_i \cap \hat{P}_i \rvert$	$\lvert N_i \cap \hat{N}_i \rvert$	$\lvert N_i \rvert$
	$\lvert \hat{P}_i \rvert$	$\lvert \hat{N}_i \rvert$	$\lvert \mathcal{L} \rvert$

From such a confusion matrix C_i, we can compute several well-known measures:

- The *Hamming loss* (HAMLOSS) computes the percentage of labels that are misclassified, i.e., relevant labels that are not predicted or irrelevant labels that are predicted. This basically corresponds to the error in the confusion matrix.

$$\text{HamLoss}(C_i) \overset{\text{def}}{=} 1 - \frac{1}{\lvert \mathcal{L} \rvert} \lvert \hat{P}_i \triangle P_i \rvert \tag{5}$$

The operator $\triangle$ denotes the symmetric difference between two sets and is defined as $A \triangle B \overset{\text{def}}{=} (A \setminus B) \cup (B \setminus A)$, i.e. $\hat{P}_i \triangle P_i$ has all labels that only appear in one of the two sets.

- *Precision* (PREC) computes the percentage of predicted labels that are relevant, *recall* (REC) computes the percentage of relevant labels that are predicted, and the F1-measure is the harmonic mean between the two.

$$\text{Prec}(C_i) \overset{\text{def}}{=} \frac{\lvert \hat{P}_i \cap P_i \rvert}{\lvert \hat{P}_i \rvert} \qquad\qquad \text{Rec}(C_i) \overset{\text{def}}{=} \frac{\lvert \hat{P}_i \cap P_i \rvert}{\lvert P_i \rvert} \tag{6}$$

$$\text{F1}(C_i) \overset{\text{def}}{=} \frac{2}{\frac{1}{\text{Rec}(C_i)} + \frac{1}{\text{Prec}(C_i)}} = \frac{2\,\text{Rec}(C_i)\,\text{Prec}(C_i)}{\text{Rec}(C_i) + \text{Prec}(C_i)} \tag{7}$$

To average these values, we compute a micro-average over all values in a test set, i.e., we add up the confusion matrices C_i for examples in the test set and compute the measure from the resulting confusion matrix. Thus, for any given measure f, the average is computed as:

$$f_{\text{avg}} = f\left(\sum_{i=1}^{m} C_i\right) \tag{8}$$

To combine the results of the individual folds of a cross-validation, we average the estimates f_{avg}^j, $j = 1 \ldots q$ over all q folds.

3.3 Perceptrons

We use the simple but fast perceptrons as base classifiers [10]. Like Support Vector Machines (SVM), their decision function describes a hyperplane that divides the N-dimensional space into two halves corresponding to positive and negative examples. We use a version that works without learning rate and threshold:

$$o'(\bar{\text{x}}) = sgn(\bar{\text{x}} \cdot \bar{\text{w}}) \tag{9}$$

with the internal weight vector $\bar{w}$ and $sgn(t) = 1$ for $t \geq 0$ and -1 otherwise. Two sets of points are called *linearly separable* if there exists a *separating hyperplane* between them. If this is the case and the examples are seen iteratively, the following update rule provably finds a separating hyperplane (cf., e.g., [11]).

$$\alpha_i = (\lambda_i - o'(\bar{x}_i)) \qquad\qquad \bar{w}_{i+1} = \bar{w}_i + \alpha_i \bar{x}_i \qquad (10)$$

It is important to see that the final weight vector can also be represented as linear combination of the training examples:

$$\bar{w} = \sum_{i=1}^{m} \alpha_i \bar{x}_i \qquad\qquad o'(\bar{x}) = sgn(\sum_{i=1}^{m} \alpha_i \cdot \bar{x}_i \bar{x}) \qquad (11)$$

assuming m to be the number of seen training examples and $\alpha_i \in \{-1, 0, 1\}$. The perceptron can hence be coded implicitly as a vector of instance weights $\alpha = (\alpha_1, \ldots, \alpha_m)$ instead of explicitly as a vector of feature weights. This representation is denominated the dual form and is crucial for developing the memory efficient variant in Section 4.

The main reason for choosing perceptrons as our base classifier is because, contrary to SVMs, they can be trained efficiently in an incremental setting, which makes them particularly well-suited for large-scale classification problems such as the Reuters-RCV1 benchmark [12], without forfeiting too much accuracy. For this reason, the perceptron has recently received increased attention (e.g. [13, 14]).

3.4   Binary Relevance Ranking

In the binary relevance (BR) or one-against-all (OAA) method, a multilabel training set with n possible classes is decomposed into n binary training sets of the same size that are then used to train n binary classifiers. So for each pair $(\bar{x}_i, P_i)$ in the original training set n different pairs of instances and binary class assignments $(\bar{x}_i, \lambda_{i_j})$ with $j = 1 \ldots n$ are generated setting $\lambda_{i_j} = 1$ if $\lambda_j \in P_i$ and $\lambda_{i_j} = -1$ otherwise. Supposing we use perceptrons as base learners, n different o'_j classifiers are trained in order to determine the relevance of λ_j. In consequence, the combined prediction of the binary relevance classifier for an instance $\bar{x}$ would be the set $\{\lambda_j \mid o'_j(\bar{x}) = 1\}$. If, in contrast, we desire a class ranking, we simply use the inner products and obtain a vector $\bar{o}(\bar{x}) = (\bar{x}\bar{w}_1, \ldots, \bar{x}\bar{w}_n)$. Ties are broken randomly to not favor any particular class.

3.5   Multiclass Multilabel Perceptrons

MMPs were proposed as an extension of the one-against-all algorithm with perceptrons as base learners [1]. Just as in binary relevance, one perceptron is trained for each class, and the prediction is calculated via the inner products. The difference lies in the update method: while in the binary relevance method all perceptrons are trained independently to return a value greater or smaller

---

Require: Training example pair $(\bar{x}, P)$, perceptrons $\bar{w}_1, \ldots, \bar{w}_n$

1: calculate $\bar{x}\bar{w}_1, \ldots, \bar{x}\bar{w}_n$, loss δ
2: **if** $\delta > 0$ **then**   ▷ only if ranking is not perfect
3:      calculate error sets E, F
4:      **for each** $\lambda \in F$ **do** $\tau_\lambda \leftarrow 0, \sigma \leftarrow 0$ ▷ initialize τ's, σ
5:      **for each** $(\lambda, \lambda') \in E$ **do**
6:          $p \leftarrow \text{PENALTY}(\bar{x}\bar{w}_1, \ldots, \bar{x}\bar{w}_n)$
7:          $\tau_\lambda \leftarrow \tau_\lambda + p$ ▷ push up pos. classes
8:          $\tau_{\lambda'} \leftarrow \tau_{\lambda'} - p$ ▷ push down neg. classes
9:          $\sigma \leftarrow \sigma + p$ ▷ for normalization
10:      **for each** $\lambda \in F$ **do**
11:          $\bar{w}_\lambda \leftarrow \bar{w}_\lambda + \delta \frac{\tau_\lambda}{\sigma} \cdot \bar{x}$ ▷ update perceptrons
12: **return** $\bar{w}_1 \ldots \bar{w}_n$ ▷ return updated perceptrons

---

Fig. 2. Pseudocode of the training method of the MMP algorithm

than zero, depending on the relevance of the classes for a certain instance, MMPs are trained to produce a good ranking so that the relevant classes are all ranked above the irrelevant classes. The perceptrons therefore cannot be trained independently, considering that the target value for each perceptron depends strongly on the values returned by the other perceptrons.

The pseudocode in Fig. 2 describes the MMP training algorithm. In summary, for each new training example the MMP first computes the predicted ranking, and if there is an error according to the chosen loss function δ (e.g. any of the losses in Sec. 3.1), it computes the set of wrongly ordered class pairs in the ranking and applies to each class in this set a penalty score according to a freely selectable function. We chose the uniform update method, where each pair in E receives the same score [1]. Please refer to [1] and [2] for a more detailed description of the algorithm.

3.6 Multilabel Pairwise Perceptrons

In the pairwise binarization method, one classifier is trained for each pair of classes, i.e., a problem with n different classes is decomposed into $\frac{n(n-1)}{2}$ smaller subproblems. For each pair of classes (λ_u, λ_v), only examples belonging to either λ_u or λ_v are used to train the corresponding classifier $o'_{u,v}$. All other examples are ignored. In the multilabel case, an example is added to the training set for classifier $o'_{u,v}$ if u is a relevant class and v is an irrelevant class, i.e., $(u, v) \in P \times N$ (cf. Figure 4). We will typically assume $u < v$, and training examples of class u will receive a training signal of $+1$, whereas training examples of class v will be classified with -1. Figure 3 shows the training algorithm in pseudocode. Of course MLPPs can also be trained incrementally.

In order to return a class ranking we use a simple voting strategy, known as *max-wins*. Given a test instance, each perceptron delivers a prediction for one of its two classes. This prediction is decoded into a vote for this particular class.

Require: Training example pair $(\bar{x}, P)$,
    perceptrons $\{\bar{w}_{u,v} \mid u < v, \lambda_u, \lambda_v \in \mathcal{L}\}$
1: **for each** $(\lambda_u, \lambda_v) \in P \times N$ **do**
2:     **if** $u < v$ **then**
3:         $\bar{w}_{u,v} \leftarrow$ TRAINPERCEPTRON$(\bar{w}_{u,v}, (\bar{x}, 1))$ ▷ train as positive example
4:     **else**
5:         $\bar{w}_{v,u} \leftarrow$ TRAINPERCEPTRON$(\bar{w}_{v,u}, (\bar{x}, -1))$ ▷ train as negative example
6: **return** $\{\bar{w}_{u,v} \mid u < v, \lambda_u, \lambda_v \in \mathcal{L}\}$ ▷ updated perceptrons

---

Fig. 3. Pseudocode of the training method of the MLPP algorithm

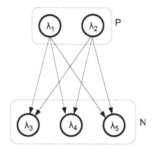

Fig. 4. MLPP training: training example $\bar{x}$ belongs to $P = \{\lambda_1, \lambda_2\}$, $N = \{\lambda_3, \lambda_4, \lambda_5\}$ are the irrelevant classes, the arrows represent the trained perceptrons

After the evaluation of all $\frac{n(n-1)}{2}$ perceptrons the classes are ordered according to their sum of votes. Ties are broken randomly in our case.

Figure 5 shows a possible result of classifying the sample instance of Figure 4. Perceptron $o'_{1,5}$ predicts (correctly) the first class, consequently λ_1 receives one vote and class λ_5 zero (denoted by $o'_{1,5} = 1$ in the first and $o'_{5,1} = -1$ in the last row). All 10 perceptrons (the values in the upper right corner can be deduced due to the symmetry property of the perceptrons) are evaluated though only six are 'qualified' since they were trained with the original example.

This may be disturbing at first sight since many 'unqualified' perceptrons are involved in the voting process: $o'_{1,2}$ is asked for instance though it cannot know anything relevant in order to determine if $\bar{x}$ belongs to λ_1 or λ_2 since it was neither trained on this example nor on other examples belonging simultaneously to both classes (or to none of both). In the worst case the noisy votes concentrate on a single negative class, which would lead to misclassifications. But note that any class can at most receive $n - 1$ votes, so that in the extreme case when the qualified perceptrons all classify correctly and the unqualified ones concentrate on a single class, a positive class would still receive at least $n - |P|$ and a negative at most $n - |P| - 1$ votes. Class λ_3 in Figure 5 is an example of this: It receives all possible noisy votes but still loses against the positive classes λ_1 and λ_2.

The pairwise binarization method is often regarded as superior to binary relevance because it profits from simpler decision boundaries in the subproblems

$o'_{1,2} = 1$	$o'_{2,1} = \text{-}1$	$o'_{3,1} = \text{-}1$	$o'_{4,1} = \text{-}1$	$o'_{5,1} = \text{-}1$
$o'_{1,3} = 1$	$o'_{2,3} = 1$	$o'_{3,2} = \text{-}1$	$o'_{4,2} = \text{-}1$	$o'_{5,2} = \text{-}1$
$o'_{1,4} = 1$	$o'_{2,4} = 1$	$o'_{3,4} = 1$	$o'_{4,3} = \text{-}1$	$o'_{5,3} = \text{-}1$
$o'_{1,5} = 1$	$o'_{2,5} = 1$	$o'_{3,5} = 1$	$o'_{4,5} = 1$	$o'_{5,4} = \text{-}1$
$v_1 = 4$	$v_2 = 3$	$v_3 = 2$	$v_4 = 1$	$v_5 = 0$

Fig. 5. MLPP voting: an example $\bar{x}$ is classified by all 10 base perceptrons $o'_{u,v}, u \neq v$, $\lambda_u, \lambda_v \in \mathcal{L}$. Note the redundancy given by $o'_{u,v} = -o'_{v,u}$. The last line counts the positive outcomes for each class.

[15, 16]. In the case of an equal class distribution, the subproblems have $\frac{2}{n}$ times the original size whereas binary relevance maintains the size. Typically, this goes hand in hand with an increase of the space where a separating hyperplane can be found. Particularly in the case of text classification the obtained benefit clearly exists. An evaluation of the pairwise approach on the Reuters-RCV1 corpus [12], which contains over 100 classes and 800,000 documents, showed a significant and substantial improvement over the MMP method [2]. This encourages us to apply the pairwise decomposition to the EUR-Lex database, with the main obstacle of the quadratic number of base classifier in relationship to the number of classes. Since this problem cannot be coped with given the present classifications in EUR-Lex we propose to reformulate the MLPP algorithm in the way described in Section 3.6.

Note that MLPP can potentially be used with any binary base learner. In particular it is possible to use advanced perceptron variants that especially consider the case of unbalanced classification problems [17, 13], since this is commonly the case for problems with a high number of classes [18]. In our opinion, this problem is not too severe for the pairwise decomposition, since it does not compare one class against the accumulation of all remaining examples, achieving on average a more balanced factor between positive and negative examples. Moreover, since we mainly evaluate the ranking quality, MMP and BR should not be discriminated by unbalanced classes.

3.7 Calibrated Label Ranking

To convert the resulting ranking of labels into a multilabel prediction, we use the *calibrated label ranking* approach [4]. This technique avoids the need for learning a threshold function for separating relevant from irrelevant labels, which is often performed as a post-processing phase after computing a ranking of all possible classes. The key idea is to introduce an artificial *calibration label* λ_0, which represents the split-point between relevant and irrelevant labels. Thus, it is assumed to be preferred over all irrelevant labels, but all relevant labels are preferred over λ_0. This introduction of an additional label during training is depicted in Figure 6, the combination with the normal pairwise base classifiers is shown in Figure 7.

As it turns out, the resulting n additional binary classifiers $\{\, o'_{0,u} \mid u = 1 \ldots n \}$ are identical to the classifiers that are trained by the binary relevance approach.

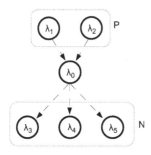

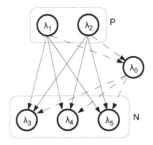

Fig. 6. Calibration: introducing virtual label λ_0 that separates P an N. Perceptrons $\bar{w}_{1,0}, \bar{w}_{2,0}, \bar{w}_{0,3}, \bar{w}_{0,4}, \bar{w}_{0,5}$ are additionally trained.

Fig. 7. CMLPP training: the complete set of trained perceptrons

Thus, each classifier $o'_{0,u}$ is trained in a one-against-all fashion by using the whole dataset with $\{\bar{x}_i \mid \lambda_u \in P_i\} \subseteq \mathcal{X}$ as positive examples and $\{\bar{x}_i \mid \lambda_u \in N_i\} \subseteq \mathcal{X}$ as negative examples. At prediction time, we will thus get a ranking over $n+1$ labels (the n original labels plus the calibration label). Then, the projection of voting aggregation of pairwise perceptrons with a calibrated label to a multilabel output is quite straight-forward:

$$\hat{P} = \{\lambda \in \mathcal{L} \mid v(\lambda) > v(\lambda_0)\}$$

where $v(\lambda)$ is the amount of votes class λ has received.

We denote the MLPP algorithm adapted in order to support the calibration technique as CMLPP. This algorithm was again applied to the large Reuters-RCV1 corpus, outperforming the binary relevance and MMP approach [4].

4   Dual Multilabel Pairwise Perceptrons

With an increasing number of classes the required memory by the MLPP algorithm grows quadratically and even on modern computers with a large memory this problem becomes unsolvable for a high number of classes. For the EU-ROVOC classification, the use of MLPP would mean maintaining approximately 8,000,000 perceptrons in memory. In order to circumvent this obstacle we reformulate the MLPP ensemble of perceptrons in dual form as we did with one single perceptron in Equation 11. In contrast to MLPP, the training examples are thus required and have to be kept in memory in addition to the associated weights, as a base perceptron is now represented as $\bar{w}_{u,v} = \sum_{i=1}^{m} \alpha_{u,v}^t \bar{x}_i$. This makes an additional loop over the training examples inevitable every time a prediction is demanded. But fortunately it is not necessary to recompute all $\bar{x}_i \bar{x}$ for each base perceptron since we can reuse them by iterating over the training examples in the outer loop, as can be seen in the following equations:

$$\bar{w}_{1,2}\bar{x} = \alpha_{1,2}^1 \bar{x}_1 \bar{x} + \alpha_{1,2}^2 \bar{x}_2 \bar{x} + \ldots + \alpha_{1,2}^m \bar{x}_m \bar{x}$$
$$\bar{w}_{1,3}\bar{x} = \alpha_{1,3}^1 \bar{x}_1 \bar{x} + \alpha_{1,3}^2 \bar{x}_2 \bar{x} + \ldots + \alpha_{1,3}^m \bar{x}_m \bar{x}$$
$$\vdots$$
$$\bar{w}_{1,n}\bar{x} = \alpha_{1,n}^1 \bar{x}_1 \bar{x} + \alpha_{1,n}^2 \bar{x}_2 \bar{x} + \ldots + \alpha_{1,n}^m \bar{x}_m \bar{x} \qquad (12)$$
$$\bar{w}_{2,3}\bar{x} = \alpha_{2,3}^1 \bar{x}_1 \bar{x} + \alpha_{2,3}^2 \bar{x}_2 \bar{x} + \ldots + \alpha_{2,3}^m \bar{x}_m \bar{x}$$
$$\vdots$$

By advancing column by column it is not necessary to repeat the dot product computations, however it is necessary to store the intermediate values, as can also be seen in the pseudocode of the training and prediction phases in Figures 8 and 9. Note also that the algorithm preserves the property of being incrementally trainable. We denote this variant of training the pairwise perceptrons the *dual multilabel pairwise perceptrons* algorithm (DMLPP).

In addition to the savings in memory and run-time, analyzed in detail in Section 5, the dual representation allows for using the kernel trick, i.e. to replace the dot product by a kernel function, in order to be able to solve originally not linearly separable problems. However, this is not necessary in our case since text problems are in general linearly separable.

Note also that the pseudocode needs to be slightly adapted when the DMLPP algorithm is trained in more than one epoch, i.e. the training set is presented to the learning algorithm more than once. It is sufficient to modify the assignment in line 8 in Figure 8 to an additive update $\alpha_{u,v}^m \leftarrow \alpha_{u,v}^m + 1$ for a revisited example $\bar{x}_m$. This setting is particularly interesting for the dual variant since, when the training set is not too big, memorizing the inner products can boost the subsequent epochs in a substantial way, making the algorithm interesting even if the number of classes is small.

Require: New training example pair $(\bar{x}_m, P_m)$,
    training examples $\bar{x}_1 \ldots \bar{x}_{m-1}$, $P_1 \ldots P_{m-1}$,
    weights $\{\alpha_{u,v}^i \mid \lambda_u, \lambda_v \in \mathcal{L}, 0 < i < m\}$
1: **for each** $\bar{x}_i = \bar{x}_1 \ldots \bar{x}_{m-1}$ **do**   ▷ iterate over previous training examples
2:     $p_i \leftarrow \bar{x}_i \cdot \bar{x}_m$
3:     **for each** $(\lambda_u, \lambda_v) \in P_m \times N_m$ **do**   ▷ $\bar{x}_m$ only relevant for training these pairs
4:         **if** $\alpha_{u,v}^i \neq 0$ **then**
5:             $s_{u,v} \leftarrow s_{u,v} + \alpha_{u,v}^i \cdot p_t$ ▷ note that $s_{u,v} = -s_{v,u}$
6: **for each** $(\lambda_u, \lambda_v) \in P_m \times N_m$ **do**   ▷ update only concerned perceptrons
7:     **if** $s_{u,v} < 0$ **then**   ▷ and only if they misspredicted
8:         $\alpha_{u,v}^m \leftarrow 1$ ▷ note that $\alpha_{u,v} = -\alpha_{v,u}$
9: **return** $\{\alpha_{u,v}^m \mid (\lambda_u, \lambda_v) \in P \times N\}$ ▷ return new weights

---

Fig. 8. Pseudocode of the training method of the DMLPP algorithm

---

Require: example $\bar{x}$ for classification,
    training examples $\bar{x}_1 \dots \bar{x}_{m-1}$, $P_1 \dots P_{m-1}$,
    weights $\{\alpha_{u,v}^i \mid \lambda_u, \lambda_v \in \mathcal{L}, 0 < i < m\}$
1: **for each** $\bar{x}_i = \bar{x}_1 \dots \bar{x}_{m-1}$ **do**   ▷ iterate over training examples
2:     $p \leftarrow \bar{x}_i \cdot \bar{x}$
3:     **for each** $(\lambda_u, \lambda_v) \in P_i \times N_i$ **do**   ▷ $\bar{x}_i$ was only be part of training these pairs
4:         **if** $\alpha_{u,v}^i \neq 0$ **then**   ▷ consider only if $\bar{x}$ is actually part of $\bar{w}_{u,v}$
5:             $s_{u,v} \leftarrow s_{u,v} + \alpha_{u,v}^i \cdot p$ ▷ add intermediate score to $\bar{w}_{u,v}\bar{x}$
6: **for each** $(\lambda_u, \lambda_v) \in \mathcal{L} \times \mathcal{L}$ **do**
7:     **if** $u \neq v \wedge s_{u,v} > 0$ **then**
8:         $v_u \leftarrow v_u + 1$ ▷ add up a vote for winning class λ_u
9: **return** voting $\bar{v} = (v_1, \dots, v_{|\mathcal{L}|})$ ▷ return voting

---

Fig. 9. Pseudocode of the prediction phase of the DMLPP algorithm

4.1 Calibration

There exist two ways of adapting the calibration approach described in Section 3.7 for DMLPP: processing the additional subproblems internally or externally.

The first version trains the additional base classifiers also in dual form. However, we believe that this approach could decrease the advantage that DMLPP obtains through the *sparseness* of the pairwise decomposition.

Therefore, the second version considered simply trains an external (non-dual) binary relevance classifier (as described in Section 3.4) in parallel. During classification, the predictions of the base perceptrons of the BR classifier are incorporated in the voting process. We will denote this algorithm as DCMLPP.

5 Computational Complexity

The notation used in this section is the following: n denotes the number of possible classes, d the average number of relevant classes per instance in the training set, N the number of attributes and N' the average number of attributes not zero (size of the sparse representation of an instance), and m denotes the size of the training set. For each complexity we will give an upper bound O in Landau notation. We will indicate the runtime complexity in terms of real value additions and multiplications ignoring operations that have to be performed by all algorithms such as sorting or internal real value operations. Additionally, we will present the complexities per instance as all algorithms are incrementally trainable. We will also concentrate on the comparison between MLPP and the implicit representation DMLPP.

The MLPP algorithm has to keep $\frac{n(n-1)}{2}$ perceptrons, each with N weights in memory, hence we need $O(n^2 N)$ memory. The DMLPP algorithm keeps the whole training set in memory, and additionally requires for each training example $\bar{x}$ access to the weights of all class pairs $P \times N$. Furthermore, it has to

Table 2. Computational complexity given in expected number of addition and multiplication operations. n: #classes, d: avg. #labels per instance, m: #training examples, N: #attributes, N': #attributes$\neq 0$.

	training time	testing time	memory requirement
MMP, BR	$O(nN')$	$O(nN')$	$O(nN)$
MLPP	$O(dnN')$	$O(n^2N')$	$O(n^2N)$
DMLPP	$O(m(dn + N'))$	$O(m(dn + N'))$	$O(m(dn + N') + n^2)$

intermediately store the resulting scores for each base perceptron during prediction, hence the complexity is $O(mdn + mN' + n^2) = O(m(dn + N') + n^2)$.[4] We can see that MLPP is applicable especially if the number of classes is low and the number of examples high, whereas DMLPP is suitable when the number of classes is high, however it does not handle huge training sets very well.

For processing one training example, $O(dn)$ dot products have to be computed by MLPP, one for each associated perceptron. Assuming that a dot product computation costs $O(N')$, we obtain a complexity of $O(dnN')$ per training example. Similarly, the DMLPP spends m dot product computations. In addition, the summation of the scores costs $O(dn)$ per training instance, leading to $O(m(dn + N'))$ operations. It is obvious that MLPP has a clear advantage over DMLPP in terms of training time, unless n is of the order of magnitude of m or the model is trained over several epochs, as already outlined in the previous Section 4.

During prediction the MLPP evaluates all perceptrons, leading to $O(n^2N')$ computations. The dual variant again iterates over all training examples and associated weights, hence the complexity is $O(m(dn + N'))$. At this phase DMLPP benefits from the linear dependence of the number of classes in contrast to the quadratic relationship of the MLPP. Roughly speaking the breaking point when DMLPP is faster in prediction is approximately when the square of the number of classes is clearly greater than the number of training documents. We can find a similar trade-off for the memory requirements with the difference that the factor between sparse and total number of attributes becomes more important, leading earlier to the breaking point when the sparseness is high. A compilation of the analysis can be found in Table 2, together with the complexities of MMP and BR. A more detailed comparison between MMP and MLPP can be found in [2].

In summary, it can be stated that the dual form of the MLPP balances the relationship between training and prediction time by increasing training and decreasing prediction costs, and especially benefits from a decreased prediction

---

[4] Note that we do not estimate d as $O(n)$ since both values are not of the same order of magnitude in practice. For the same reason we distinguish between N and N' since particularly in text classification both values are not linked: a text document often turns out to employ around 100 different words whereas the size of the vocabulary of a the whole corpus can easily reach 100,000 words (although this number is normally reduced by feature selection).

time and memory savings when the number of classes is large. Thus, this technique addresses the main obstacle to applying the pairwise approach to problems with a large number of labels.

For the complexities of the calibrated variants of MLPP and DMLPP we can simply add the corresponding complexity of BR, at least if we consider the externally calibrated variant of DCMLPP.

6   Experiments

For the MMP algorithm we used the IsErr loss function and the uniform penalty function. This setting showed the best results in [1] on the RCV1 data set. The perceptrons of the BR and MMP ensembles were initialized with random values. We performed also tests with a multilabel variant of the multinomial Naive Bayes (MLNB) algorithm in order to provide a baseline. Another baseline is depicted by FC (frequency classifier) that returns always the same ranking of classes according to the class frequency in the training set.

6.1   Ranking Quality

The results for the four algorithms and the three different classifications of EUR-Lex are presented in Table 3. DMLPP results are omitted since they differ only slightly from those of DCMLPP due to the possible additional (one) vote won against the artificial label. In the same way, we omit the results of MLPP since they differ only marginally due to a different random initialization. Note however that MLPP cannot be applied to the *EUROVOC* dataset due to the high memory requirements, which was the reason for developing the dual version.

The values for IsErr, OneErr, RankLoss and AvgP are shown ×100% for better readability, AvgP is also presented in the conventional way (with 100% as the optimal value) and not as a loss function. The number of epochs indicates the number of times that the online-learning algorithms were able to see the training instances. No results are reported for the performance of DCMLPP on *EUROVOC* for more than two epochs due to time restrictions. Note also that the results differ slightly from those of previous experiments in [6, 7] due to the modifications to the dataset presented in Section 2.

The first appreciable characteristic is that DCMLPP dominates all other algorithms on all three views of the EUR-Lex data, regardless of the number of epochs or losses. Often DCMLPP achieves better results than the other algorithms for more epochs. Especially on the losses that directly evaluate the ranking performance the improvement is quite pronounced and the results are already unreachable after the first epoch.

In addition to the fact that the DMLPP outperforms the remaining algorithms, it is still interesting to compare the performances of MMP and BR as they have still the advantage of reduced computational costs and memory requirements in comparison to the (dual) pairwise approach and could therefore be more applicable for very complex data sets such as *EUROVOC*, which is certainly hard to tackle for DMLPP (cf. Section 6.3).

Table 3. Average ranking losses for the three views on the data and for the different algorithms. For IsErr, OneErr, RankLoss and Margin low values are good, for AvgP the higher the better.

			1 epoch			2 epochs			5 epochs			10 epochs		
	FC	MLNB	BR	MMP	DCMLPP	BR	MMP	DCMLPP	BR	MMP	DCMLPP	BR	MMP	DCMLPP
subject matter														
IsErr ×100	99.58	99.47	65.99	55.70	51.38	58.78	51.96	44.07	53.42	42.77	38.23	50.19	40.22	36.34
OneErr ×100	77.83	98.68	35.71	30.58	22.78	27.13	27.09	17.29	22.69	18.38	13.49	20.64	15.97	12.55
RankLoss	12.89	8.885	17.38	2.303	1.064	13.89	2.520	0.911	11.58	2.091	0.796	9.752	1.85	0.762
Margin	40.16	25.04	62.31	10.11	4.316	52.28	11.22	3.757	44.77	9.366	3.337	38.45	8.177	3.214
AvgP	22.57	11.91	59.33	74.01	78.68	66.07	76.95	82.73	70.69	82.10	85.64	73.30	83.75	86.52
directory code														
IsErr ×100	91.51	99.34	52.80	47.68	36.55	46.26	40.01	32.38	40.76	33.28	29.22	37.55	31.39	28.30
OneErr ×100	90.13	99.04	44.40	40.85	28.22	37.38	32.99	24.42	31.48	25.79	21.41	28.1	23.9	20.65
RankLoss	14.17	7.446	19.40	2.383	0.972	15.09	2.058	0.863	11.69	1.874	0.824	9.876	1.529	0.815
Margin	68.33	34.44	96.43	14.18	5.626	77.32	12.18	5.045	61.48	10.95	4.831	52.94	8.947	4.785
AvgP	18.98	6.714	57.10	68.70	77.89	63.68	74.90	80.87	68.75	79.84	82.87	71.61	81.30	83.38
EUROVOC														
IsErr ×100	99.82	99.82	99.25	99.14	98.20	98.70	98.00	96.75	97.46	96.14		97.06	95.13	
OneErr ×100	93.52	99.58	53.11	78.98	34.76	44.93	56.88	28.01	36.69	39.46		33.84	34.99	
RankLoss	12.97	22.34	39.78	3.669	2.692	35.25	4.091	2.398	30.93	4.573		28.59	4.509	
Margin	1357.10	1623.72	3218.12	562.81	426.28	3040.01	670.65	387.51	2846.47	757.01		2716.63	740.12	
AvgP	5.504	1.060	25.55	27.04	46.79	30.71	38.42	52.72	35.95	47.65		38.31	50.71	

Table 4. Computational costs in CPU-time and millions of real value operations (M op.)

subject matter	training		testing	
BR	29.96 s	1,680 M op.	7.09 s	184 M op.
MMP	31.95 s	1,807 M op.	6.89 s	184 M op.
DMLPP	372.14 s	6,035 M op.	151.98 s	4,471 M op.
MLPP	69.50 s	3,886 M op.	164.04 s	18,427 M op.

directory code	training		testing	
BR	50.67 s	3,420 M op.	9.62 s	378 M op.
MMP	53.38 s	3,615 M op.	9.46 s	378 M op.
DMLPP	383.40 s	3,047 M op.	187.65 s	5,246 M op.
MLPP	120.70 s	4,735 M op.	643.34 s	77,629 M op.

EUROVOC	training		testing	
BR	368.02 s	33,074 M op.	53.34 s	3,662 M op.
MMP	479.14 s	40,547 M op.	52.90 s	3,662 M op.
DMLPP	13,058.01 s	17,647 M op.	6,780.51 s	123,422 M op.
MLPP	–		–	

For the *subject matter* and *directory code*, the results clearly show that the MMP algorithm outperforms the simple one-against-all approach. Especially on the losses that directly evaluate the ranking performance the improvement is quite pronounced. The smallest difference can be observed in terms of ONEERR, which evaluates the top class accuracy.

The performance on the *EUROVOC* descriptor data set confirms the previous results. The differences in RANKLOSS and MARGIN are very pronounced. In contrast, in terms of ONEERR the MMP algorithm is worse than one-against-all, even after ten epochs. It seems that with an increasing amount of classes, the MMP algorithm has more difficulties to push the relevant classes to the top such that the margin is big enough to leave all irrelevant classes below, although the algorithm in general clearly gives the relevant classes a higher score than the one-against-all approach. An explanation could be the dependence between the perceptrons of the MMP. This leads to a natural normalization of the scalar product, while there is no such restriction when trained independently as done in the binary relevance algorithm. As a consequence there could be some perceptrons that produce high maximum scores and thereby often arrive at top positions at the overall ranking. Furthermore, MMP's accuracy on RANKLOSS and MARGIN seems to suffer from the increased number of classes, since the loss increases from the first to the fifth epoch, and still in the tenth epoch the value is higher than after only one epoch. Perhaps it is indicated to use a different loss for MMP to optimize for problems with higher amount of classes, where ISERR is inevitably high (cf. Section 3.5). The price to pay for the good ONEERR of BR is a decreased quality of the produced rankings, as the results for RANKLOSS and MARGIN are even beaten by Naive Bayes, which is by far the worst algorithm for the other losses.

It is interesting to note in this context that the frequency classifier often achieves a better performance than Naive Bayes and even BR, especially with increasing number of classes as with *EUROVOC*.

The fact that in only approximately 5% of the cases a perfect classification is achieved and in only approx. 65% the top class is correctly predicted in *EUROVOC* (MMP) should not lead to an underestimation of the performance of these algorithms. Considering that with almost 4000 possible classes and only 5.3 classes per example the probability of randomly choosing a correct class is less than one percent, namely 0.13%, the performance is indeed substantial.

6.2   Multilabel Classification Quality

Table 5 shows the several results for predicting a set of labels for each instance rather than a ranking of labels. Obviously, only results for BR and the calibrated version of DMLPP can be shown since MMP only produces a ranking. The first remarkable point is that DCMLPP outperforms BR in all direct comparisons for the overall measures HAMLOSS and F1 and also PREC. But interestingly, BR always achieves a higher REC than DCMLPP. This is due to the fact that the calibration tends to underestimate the number of returned labels for each instance, especially for a high number of total classes and when the base classifiers are not yet that accurate such as for low numbers of epochs. A possible

Table 5. Average multilabel losses for the three views on the data and for the label set predicting BR and DCMLPP. For HAMLOSS low values are good, for the remaining measures high values near 100% are good.

		1 epoch		2 epochs		5 epochs		10 epochs	
		BR	DCMLPP	BR	DCMLPP	BR	DCMLPP	BR	DCMLPP
subject matter	HAMLOSS	1.196	0.715	1.004	0.641	0.823	0.574	0.757	0.540
	F1	54.39	62.43	60.13	68.81	65.73	72.66	68.32	74.47
	REC	64.64	54.02	68.66	64.26	71.62	69.25	74.11	71.56
	PREC	47.03	74.08	53.55	74.09	60.74	76.43	63.39	77.63
directory code	HAMLOSS	0.416	0.231	0.355	0.198	0.289	0.179	0.265	0.169
	F1	46.81	49.37	53.28	62.95	59.74	67.75	62.58	69.64
	REC	58.31	36.05	64.51	53.55	68.36	59.87	70.54	61.89
	PREC	39.13	78.56	45.41	76.38	53.07	78.04	56.26	79.61
EUROVOC	HAMLOSS	0.267	0.125	0.238	0.117	0.208		0.199	
	F1	26.95	18.20	31.56	36.11	36.42		38.57	
	REC	37.03	10.45	41.30	24.89	44.84		46.93	
	PREC	21.19	71.62	25.54	65.82	30.67		32.74	

explanation for this behavior is the following: when the BR classifier, that is also included in DCMLPP or CMLPP, predicts that v classes are positive, this means for the remaining classes that they have to obtain at least $n - v$ votes of their maximum of n votes in order to be predicted as positive. The probability that this happens for a real positive class decreases with increasing n and increasing error of the base classifiers, since it becomes more probably that at least v base classifiers mistakenly commit a wrong decision.

The average label set size that is produced by DCMLPP demonstrates this: for *subject matter* it increases from 1.65 to 2.04 (BR from 3.05 to 2.59), for *directory code* from 0.59 to 1.0 (BR: 1.93 to 1.62) and for *EUROVOC* it increases from small 0.77 to 2.01 after the second epoch (BR from 9.28 to 7.61 in the tenth epoch). BR begins with an overestimation, reducing the predicted size subsequently.

In order to allow a comparison independent of different tendencies of the different thresholding techniques, we have computed REC and PREC using the correct, true size of the label set of the test examples. With this trick, we obtained a REC/PREC of 68.5% for BR and 79.3 for DCMLPP on the last epoch of *subject matter*. Note that REC equals always PREC since we have always the same amount of false positives and false negatives in the confusion matrix. For *directory code* the values are 64.8 and 75.0, and for *EUROVOC* 33.6 and 48.0 for the second epoch, 40.8 for the tenth epoch for BR. This trick can also be applied in order to compare to MMP, which is beaten by DCMLPP but better than BR with 76.2, 71.9, 35.2 and 47.1 respectively.

6.3 Computational Costs

In order to allow a comparison independent from external factors such as logging activities and the run-time environment, we ignored minor operations that have to be performed by all algorithms, such as sorting or internal operations. An overview over the amount of real value addition and multiplication computations is given in Table 4 (averaged over the cross validation splits, trained for one epoch), together with the CPU-times on an AMD Dual Core Opteron 2000 MHz as additional reference information. We report only results of DMLPP, since DCMLPP's operations and seconds can easily be derived or estimated by adding those of BR. Furthermore, we include the results for the non-dual MLPP, however no values have been received for the *EUROVOC* problem due to the memory space problem discussed at the end of this section.

We can observe a clear advantage of the non-pairwise approaches on the *subject matter* data especially for the prediction phase, however the training costs are in the same order of magnitude. Between MLPP and DMLPP we can see an antisymmetric behavior: while MLPP requires only almost half of the amount of the DMLPP operations for training, DMLPP reduces the amount of prediction operations by a factor of more than 4. For the *directory code* the rate for MMP and BR more than doubles in correspondence with the increase in number of classes, additionally the MLPP testing time substantially increases due to the quadratic dependency, while DMLPP profits from the decrease in the average number of classes per instance. It even causes less computations in the training phase than MMP/BR. The reason for this is not only the reduced maximum amount of weights per instance (cf. Section 5), but particularly the decreased probability that a training example is relevant for a new training example (and consequently that dot products and scores have to be computed) since it is less probable that both class assignments match, i.e. that both examples have the same pair of positive and negative classes. This becomes particularly clear if we observe the number of non-zero weights and actually used weights during training for each new example. The classifier for *subject matter* has on average 20 weights set per instance out of 440 ($= d(n-d)$) in the worst case (a ratio of 4.45%), and on average 4.97% of them are required when a new training example arrives. For the *directory code* with a smaller fraction d/n 35.0 weights are stored (6.66%), of which only 1.10% are used when updating. This also explains the relatively small number of operations for training on *EUROVOC*, since from the 1,781 weights per instance (8.45%), only 0.55% are relevant to a new training instance. In this context, regarding the disturbing ratio between real value operations and CPU-time for training DMLPP on *EUROVOC*, we believe that this is caused by a suboptimal storage structure and processing of the weights and we are therefore confident that it is possible to reduce the distance to MMP in terms of actual consumed CPU-time by improving the program code. Memory swapping may also have influenced the measurement.

Note that MMP and BR compute the same amount of dot products, the computational costs only differ in the number of vector additions, i.e. perceptron updates. It is therefore interesting to observe the contrary behavior of both algorithms when the number of classes increases: while the one-against-all algorithm reduces the ratio of updated perceptrons per training example from 1.33% to 0.34% when going from 202 to 3993 classes, the MMP algorithm more than

Table 6. Memory requirements of the different classifiers for the EUR-Lex datasets

dataset	BR/MMP	DMLPP	DCMLPP	MLPP
subject matter	153 MB	199 MB	210 MB	541 MB
directory code	167 MB	210 MB	229 MB	1,818 MB
EUROVOC	1,145 MB	1,242 MB	1,403 MB	–

doubles the rate from 8.53% to 22.22%. For the MMP this behavior is natural: with more classes the error set size increases and consequently the number of updated perceptrons. In contrast BR receives less positive examples per base classifier, the perceptrons quickly adopt the generally good rule to always return a negative score, which leads to only a few binary errors and consequently to little corrective updates. A more extensive comparison of BR and MMP can be found in previous work [19].

The memory consumption provided by the Java Virtual Machine after training the several classifiers for one epoch is depicted in Table 6. Note that these sizes include the overhead caused by the virtual machine and the machine learning framework.[5] MLPP already consumes more memory than the dual variant for the first dataset with 200 classes. For the 400 classes of the *directory code* view the algorithm requires almost 2 GB, while DMLPP is able to compress the same information into slightly more than 200 MB. As expected and already mentioned in Section 6.1, MLPP is not applicable to *EUROVOC*. A simple estimation based on the number of base classifiers, number of features and bytes per float variable results in 152 GB of memory. Another remarkable fact is that the memory requirement of DMLPP is comparable to that of the one-per-class algorithms: for the smaller datasets we obtain an overhead of only 50 MB and for the bigger *EUROVOC* view it requires only double of the memory, although representing a quadratic number of base classifiers. On the other hand, a view on the memory consumption of DCMLPP reveals that great part of the space for MMP/BR and DCMLPP is caused by overhead of the JVM and the machine learning framework (instances and class mappings in memory, extensive statistics, etc.). If we compute the core memory requirements of BR by subtracting DMLPP's value from DMLPP's for *EUROVOC*, we obtain 161 MB. In consequence we obtain a general overhead of 981 MB and thus an actual memory consumption of 261 MB for DMLPP. These values seem to be realistic after a simple estimation.

7   Conclusions

In this paper, we introduced the EUR-Lex text collection as a promising test bed for studies in text categorization, available at

---

[5] We used the WEKA framework (http://www.cs.waikato.ac.nz/~ml/weka/), but we adapted it so that it maintains a copy of a training instance in memory only when necessary for the incremental updating.

http://www.ke.tu-darmstadt.de/resources/eurlex/. Among its many interesting characteristics (e.g., multi-linguality), our main interest was the large number of categories, which is one order of magnitude above other frequently studied text categorization benchmarks, such as the Reuters-RCV1 collection.

On the *EUROVOC* classification task, a multilabel classification task with 4000 possible labels, the DMLPP algorithm, which decomposes the problem into training classifiers for each pair of classes, achieves an average precision rate of slightly more than 50%. Roughly speaking, this means that the (on average) five relevant labels of a document will (again, on average) appear within the first 10 ranks in the relevancy ranking of the 4,000 labels. This is a very encouraging result for a possible automated or semi-automated real-world application for categorizing EU legal documents into EUROVOC categories.

This result was only possible by finding an efficient solution for storing the approx. 8,000,000 binary classifiers that have to be trained by this pairwise approach. To this end, we showed that a reformulation of the pairwise decomposition approach into a dual form is capable of handling very complex problems and can therefore compete with the approaches that use only one classifier per class. It was demonstrated that decomposing the initial problem into smaller problems for each pair of classes achieves higher prediction accuracy on the EUR-Lex data, since DMLPP substantially outperforms all other algorithms. This confirms previous results of the non-dual variant on the large Reuters Corpus Volume 1 [2]. The dual form representation allows for handling a much higher number of classes than the explicit representation, albeit with an increased dependence on the training set size. Despite the improved ability to handle large problems, DMLPP is still less efficient than MMP, especially for the *EUROVOC* data with 4000 classes. However, in our opinion the results show that DMLPP is still competitive for solving large-scale problems in practice, especially considering the trade-off between runtime and prediction performance.

As further work, we have adapted the efficient voting technique for pairwise classification introduced in [20] to the multilabel case. This technique permits to reduce the amount of perceptron predictions of the MLPP algorithm during the classification of the *subject matter* and *directory code* views to a level competitive to BR/MMP [21]. However, the processing of the almost 4000 classes of *EUROVOC* is out of scope, making it still necessary to use techniques such as the Dual MLPP.

Additionally, we are currently investigating hybrid variants to further reduce the computational complexity. The idea is to use a different formulation in training than in the prediction phase depending on the specific memory and runtime requirements of the task. In order e.g. to combine the advantage of MLPP during training and DMLPP during predicting on the *subject matter* subproblem, we could train the classifier as in the MLPP (with the difference of iterating over the perceptrons first so that only one perceptron has to remain in memory at a point in time) and then convert it to the dual representation by means of the collected information during training the perceptrons. The use in training of SVMs or more advanced perceptron variants that, similar to SVMs, try to maximize the margin of the separating hyperplane in order to produce more accurate models [13, 14], is also an interesting option.

Acknowledgements. This work was supported by the EC 6th framework project ALIS (Automated Legal Information System) and by the German Science Foundation (DFG).

References

[1] Crammer, K., Singer, Y.: A Family of Additive Online Algorithms for Category Ranking. Journal of Machine Learning Research 3, 1025–1058 (2003)

[2] Loza Mencía, E., Fürnkranz, J.: Pairwise learning of multilabel classifications with perceptrons. In: Proceedings of the 2008 IEEE International Joint Conference on Neural Networks (IJCNN 2008), Hong Kong, pp. 2900–2907 (2008)

[3] Brinker, K., Fürnkranz, J., Hüllermeier, E.: A Unified Model for Multilabel Classification and Ranking. In: Proceedings of the 17th European Conference on Artificial Intelligence, ECAI 2006 (2006)

[4] Fürnkranz, J., Hüllermeier, E., Loza Mencía, E., Brinker, K.: Multilabel classification via calibrated label ranking. Machine Learning 73, 133–153 (2008)

[5] Sebastiani, F.: Machine learning in automated text categorization. ACM Computing Surveys 34, 1–47 (2002)

[6] Loza Mencía, E., Fürnkranz, J.: Efficient multilabel classification algorithms for large-scale problems in the legal domain. In: Proceedings of the Language Resources and Evaluation Conference (LREC) Workshop on Semantic Processing of Legal Texts, Marrakech, Morocco, pp. 23–32 (2008)

[7] Loza Mencía, E., Fürnkranz, J.: Efficient pairwise multilabel classification for large-scale problems in the legal domain. In: Daelemans, W., Goethals, B., Morik, K. (eds.) Proceedings of the European Conference on Machine Learning and Principles and Practice of Knowledge Disocvery in Databases (ECML-PKDD 2008), Part II, Antwerp, Belgium, pp. 50–65. Springer, Heidelberg (2008)

[8] Yang, Y., Pedersen, J.O.: A comparative study on feature selection in text categorization. In: ICML 1997: Proceedings of the Fourteenth International Conference on Machine Learning, pp. 412–420. Morgan Kaufmann Publishers Inc., San Francisco (1997)

[9] Pouliquen, B., Steinberger, R., Ignat, C.: Automatic annotation of multilingual text collections with a conceptual thesaurus. In: Proceedings of the Workshop Ontologies and Information Extraction at the Summer School, The Semantic Web and Language Technology - Its Potential and Practicalities (EUROLAN 2003), Bucharest, Romania, July 28 - August 8 (2003)

[10] Rosenblatt, F.: The perceptron: a probabilistic model for information storage and organization in the brain. Psychological Review 65, 386–408 (1958)

[11] Bishop, C.M.: Neural Networks for Pattern Recognition. Oxford University Press, Oxford (1995)

[12] Lewis, D.D., Yang, Y., Rose, T.G., Li, F.: RCV1: A New Benchmark Collection for Text Categorization Research. Journal of Machine Learning Research 5, 361–397 (2004)

[13] Crammer, K., Dekel, O., Keshet, J., Shalev-Shwartz, S., Singer, Y.: Online passive-aggressive algorithms. Journal of Machine Learning Research 7, 551–585 (2006)

[14] Khardon, R., Wachman, G.: Noise tolerant variants of the perceptron algorithm. Journal of Machine Learning Research 8, 227–248 (2007)
[15] Fürnkranz, J.: Round Robin Classification. Journal of Machine Learning Research 2, 721–747 (2002)
[16] Hsu, C.W., Lin, C.J.: A Comparison of Methods for Multi-class Support Vector Machines. IEEE Transactions on Neural Networks 13, 415–425 (2002)
[17] Li, Y., Zaragoza, H., Herbrich, R., Shawe-Taylor, J., Kandola, J.S.: The Perceptron Algorithm with Uneven Margins. In: Proceedings of the Nineteenth International Conference on Machine Learning (ICML 2002), pp. 379–386 (2002)
[18] Montejo Ráez, A., Ureña López, L.A., Steinberger, R.: Adaptive selection of base classifiers in one-against-all learning for large multi-labeled collections. In: Vicedo, J.L., Martínez-Barco, P., Muñoz, R., Saiz Noeda, M. (eds.) EsTAL 2004. LNCS (LNAI), vol. 3230, pp. 1–12. Springer, Heidelberg (2004)
[19] Loza Mencía, E., Fürnkranz, J.: An evaluation of efficient multilabel classification algorithms for large-scale problems in the legal domain. In: LWA 2007: Lernen - Wissen - Adaption, Workshop Proceedings, pp. 126–132 (2007)
[20] Park, S.H., Fürnkranz, J.: Efficient pairwise classification. In: Kok, J.N., Koronacki, J., Lopez de Mantaras, R., Matwin, S., Mladenič, D., Skowron, A. (eds.) ECML 2007. LNCS (LNAI), vol. 4701, pp. 658–665. Springer, Heidelberg (2007)
[21] Loza Mencía, E., Park, S.H., Fürnkranz, J.: Efficient voting prediction for pairwise multilabel classification. In: Proceedings of the 11th European Symposium on Artificial Neural Networks (ESANN 2009). Springer, Heidelberg (2009)

An Automatic System for Summarization and Information Extraction of Legal Information

Emmanuel Chieze[1,*], Atefeh Farzindar[2], and Guy Lapalme[3]

[1] Département d'informatique
Université du Québec à Montréal
C.P. 8888, Succ. Centre-ville
Montréal, Québec, Canada, H3C 3P8
chieze.emmanuel@uqam.ca
[2] *NLP Technologies Inc*
3333, chemin Queen Mary, suite 543
Montréal, Québec, Canada H3V 1A2
farzindar@nlptechnologies.ca
[3] RALI-DIRO
C.P. 6128, Succ. Centre-ville
Université de Montréal
Montréal, Québec, Canada H3C 3J7
lapalme@iro.umontreal.ca

Abstract. This paper presents an information system for legal professionals that integrates natural language processing technologies such as text classification and summarization. We describe our experience in the use of a mix of linguistics aware transductor and XML technologies for bilingual information extraction from judgements in both French and English within a legal information and summarizing system. We present the context of the work, the main challenges and how they were tackled by clearly separating language and domain dependent terms and vocabularies. After having been developed on the immigration law domain, the system was easily ported to the intellectual property and tax law domains.

Keywords: Summarization, Natural Language Processing, Information Extraction.

1 Context of the Work

Legal information is produced in large quantities and it needs to be adequately classified in order to be reliably accessible. Indeed, legal experts perform relatively difficult legal clerical work that requires accuracy and speed. These legal experts often summarize legal documents, such as court judgements, and look for information relevant to specific cases in these summaries. These tasks involve

---

* This work was performed while Emmanuel Chieze was at RALI-DIRO, Université de Montréal.

E. Francesconi et al. (Eds.): Semantic Processing of Legal Texts, LNAI 6036, pp. 216–234, 2010.
© Springer-Verlag Berlin Heidelberg 2010

understanding, interpreting, explaining and researching a wide variety of legal documents.

To help in some of these tasks, *NLP Technologies*[1] has developed a series of advanced information technologies in the judicial domain. *NLP Technologies* is an automated language software company conducting the research, development and marketing of summarization and statistical machine translation software and related software tools and services. The company's services are available through the company's website and include access to four main tools: *DecisionExpress*, *SearchExpress*, *BiblioExpress* and *StatisticExpress* which are briefly described below.

The core technology underlying these tools is an automatic summarization system. Summaries help organize large volumes of documents so that finding relevant judgements for a specific case is both easy and efficient. That is why judgements are frequently manually summarized by legal experts. However, human time and expertise required to provide manual summaries for legal research make human-generated summaries relatively expensive. Also, there is always the risk that a legal expert misinterprets a judgement and misclassifies it or produces an erroneous summary. Because of the high accuracy required in the classification and summarization of legal judgements, commonly available automatic classification and summarization methods are typically not suitable for this task. Based on the work of Farzindar [2], *NLP Technologies* has developed a summarization system specifically tailored for the legal domain based on a thematic segmentation of the text. Since 2005, the Federal Court of Canada has been a client of *NLP Technologies*'s automated legal analysis services for French and English documents. The summaries are available within 2 days of the publication date. Although this process must be adapted for new domains, the fundamentals stay the same and one of the goals of this work was to develop a methodology that allows an easy parameterization process through appropriate dictionaries and rules using advanced natural language processing tools such as transductors.

1.1 *DecisionExpress*

DecisionExpress is a weekly bulletin of recent decisions of Canadian federal courts and provincial tribunals. It processes judicial decisions automatically and makes the daily information used by jurists more accessible by presenting the summaries of the legal record of the proceedings of federal courts (such as Tax court, Federal court of appeal, etc.) and provincial tribunals in Canada.

Furthermore, it presents a factsheet for each decision that can save hours of reading by extracting the essential information and showing it in a user-friendly format for many cases of the same type.

Contrary to the traditional way of manually classifying and summarizing judgements to be saved in the database, *DecisionExpress* analyses and summarizes the judgements automatically. This brings numerous advantages both for those publishing legal information and the jurists using it:

---

[1] http://www.nlptechnologies.ca

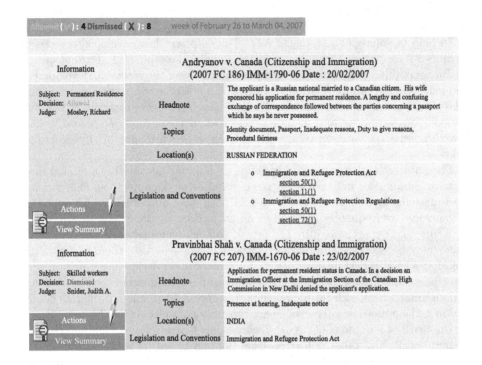

Fig. 1. Factsheet from *DecisionExpress* showing two cases from a week in which 4 immigration cases have been allowed and 8 dismissed. The left part gives the subject, the decision and the name of the judge while the right part gives a very short summary, the topics dealt with in this case, the country in which the applicant resided and the pertinent legislation that was cited in the case. Clicking on the appropriate button gives access to a longer summary (Figure 2) or the text of the original judgement.

- Significant cost reduction of the summary production process which can be passed back to those accessing the information as customers.
- Automatic summaries present sentences extracted from the judgement, whereas manual summaries consist of reformulations. A reformulation is less precise and less credible because it is not a direct source of law. In addition, an ambiguous reformulation can may lead to misinterpretations of what the judge meant and lead the user to erroneous beliefs.
- Automatic summaries provide greater consistency. The editors' abilities and concentration may vary, whereas the computer provides a stable level of performance. The machine is also better suited than a human for repetitive tedious tasks, such as the production of summaries of long articles.

DecisionExpress' other innovation is the production of a brief description of every decision analysed. This description allows a jurist to get the essential information of a decision in one glance. This way, he or she knows immediately if the decision is relevant enough to read the summary and eventually the whole judgement.

The thematic segmentation is based on specific knowledge of the legal field. According to our analysis, legal texts have a thematic structure independent of the category of the judgement [1]. Textual units dealing with the same subject form a thematic segment set. In this context, we distinguish four themes, which divide the legal decisions into thematic segments, based on the work of judge Mailhot[5]:

Introduction describes the situation before the court and answers these questions: who did what to whom?

Context explains the facts in chronological order: it describes the story including the facts and events related to the parties and it presents findings of credibility related to the disputed facts.

Reasoning describes the comments of the judge and the finding of facts, and the application of the law to the found facts. This section of the judgement is the most important part for legal experts because it presents the solution to the problem between the parties and leads the judgement to a conclusion.

Conclusion expresses the disposition, which is the final part of a decision containing the information about what is decided by the court.

The factsheet (see Figure 1) presents information such as the name of the judge who signed the judgement and the tribunal he or she belongs to, the domain of law and the subject of the decision (for example, immigration and application for permanent residence), a short description of the litigated point, the judge's conclusion (allowed or dismissed) and hyperlinks to the summary (Figure 2) and the original judgement.

These factsheets are highly appreciated by users because they present the essential information about a judgement more concisely than a summary. One glance is enough to determine if the decision is relevant. Moreover, the factsheets are automatically translated into French or English so that for every decision, the factsheet is available in both offical languages of Canada. This allows jurists to work in the language they are most comfortable with regardless of the language in which the decision was published.

1.2 *SearchExpress*

SearchExpress, integrated within *DecisionExpress* is a search engine that allows users to search the *NLP Technologies'* database rendered by Canadian federal courts and tribunals. In addition to the search functionality already offered by most providers of legal information such as QuickLaw[2] and Westlaw-Carswell[3], *SearchExpress* offers new possibilities. Search the factsheets generated by *DecisionExpress*. This way, the user can formulate the query based on the judge's name, his conclusion, the domain of law, the subject of the decision, the keywords, etc. In short, the query can be constructed using any information presented in the factsheets, which allows the user to refine his or her search.

---

[2] http://www.lexisnexis.ca

[3] http://www.carswell.com

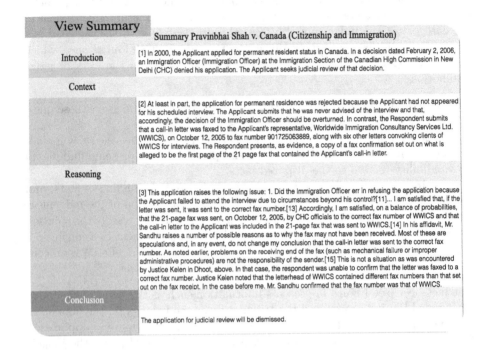

Fig. 2. Automatically generated, and manually reviewed, summary returned after clicking on the View Summary button at the bottom left of Figure 1. All sentences of the summary being taken verbatim from the original decision, they can thus be used more easily in legal argument. The sentences are classified into meaningful sections: Introduction, Context, Reasoning and Conclusion. Note that sentences are not necessarily in the same order in the judgement and in the summary.

Regardless of the type of search used, the results page presents, for every decision found, the factsheet of the decision as well as a hyperlink to the original text. This manner of presenting the results permits the user to save time in the preliminary sort of retrieved results. Instead of reading every decision retrieved to see if it is relevant, he or she can simply reject the decisions whose factsheets show clearly that they will not be useful. The overview presented in the factsheets also allows telling quickly if the query should be refined or otherwise modified.

Searching is done both in the full text of the judgement and in the factsheets generated by *DecisionExpress*. Consider a lawyer preparing a file for a client contesting in the Federal Court the refusal of his application for residence based on humanitarian considerations. On the other providers' websites, this lawyer could do a search in the decisions of the Federal Court by typing keywords such as immigration or application for refugee status (very broad) or humanitarian considerations (more precise but not always related specifically to the humanitarian application process per se). With *SearchExpress*, the lawyer can search only those cases which have been correctly identified as judicial reviews of humanitarian applications per se among the Federal Court by limiting the results of the query to judgements labelled by *DecisionExpress* as immigration for the domain, and "humanitarian and compassionate application" for the subject. He or she could also choose only to retrieve decisions where the judge has granted the application for judicial review, or else limit the search to decisions with respect to applicants from the same country as his or her client. Cross-lingual (English-French) search allows the user to submit a query in one language and retrieve documents containing the terms of the query as well as their equivalents in the other language. The user can thus use a single query to retrieve all documents relevant to his or her case regardless of the language the judgement was made in. *SearchExpress* makes search easier by incorporating unique and useful search criteria (Figure 3) such as category (e.g., immigration and tax), court, name of judge, subject (e.g., investors, pre-removal risk assessment, and humanitarian considerations), conclusion (allowed or dismissed) and other relevant criteria.

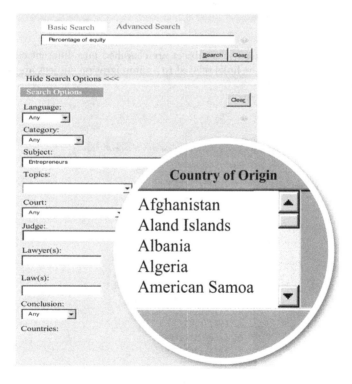

Fig. 3. With *SearchExpress*, it is possible to have access to relevant decisions for a research in progress by specifying the criteria. For example, a lawyer can carry out research on a precise case such as an entrepreneur who comes from a certain country with particular conditions to see how such situations have been treated historically in order to calculate the chance of success in court.

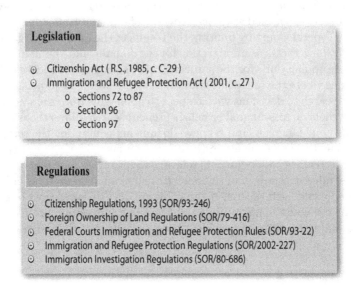

Fig. 4. The legal sources related to a subject are classified into different categories by *BiblioExpress*. For example, the links related to an immigration subject are centralized in one page.

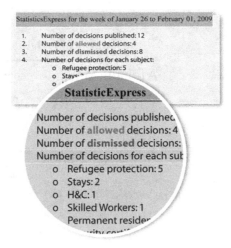

Fig. 5. Weekly statistics provided by *StatisticExpress*

1.3 BiblioExpress

BiblioExpress provides access to the text of federal legislations, rules, policy manuals and guidelines, as well as a range of inter-governmental agreements and international instruments in Canada. This service (see Figure 4) centralizes links to fundamental legal resources of three different categories: immigration, intellectual property, and tax. For instance, in the *Immigration of Canada* domain, *BiblioExpress* classifies the links to recourses into legislations, regulations, rules, conventions, guidelines, forms, agreements, etc.

1.4 StatisticExpress

StatisticExpress gives access to pertinent data and a variety of government statistics such as the annual and periodical reports of courts, tribunals and government agencies, international statistics, the performance reports of various government and international agencies and a specialized fact-finder providing statistics from *DecisionExpress*'s databases shown in Figure 5.

2   Research Background

The best source for an overview of legal text summarization is Moens [6] who presents an excellent survey of the area of summarization of court decisions. She describes the context in which court decisions are taken and published and the need for good quality summaries in this area which is comparable to the medical domain.

FLEXICON [9] is one of the first summarization system specialized for legal texts. It was based on the use of keywords found in a legal phrase dictionary. The summaries were not used as such but served for indexing a legal case text collection. SALOMON [7], developed for summarization of Belgian criminal cases, was the first to explicitly make use of the structure of a case. As such, the authors were more interested in identifying the structure than producing a complete summary. The SUM [4] project was developed to determine the rhetorical status of sentences of House of Lords judgements. This methodology could be used as a background technology for a complete summarization system.

These projects attest to the importance of the exploration of legal knowledge for sentence categorisation and summarization. *NLP Technologies*'s extraction of the most important units is based on the identification of the thematic structure in the document and the determination of argumentative themes of the textual units in the judgement[1,2]. The system we describe in this paper is the only one that spans all steps from an original judgement to a complete summaries that can be used in the daily activity of legal professionals.

3   The Immigration and Refugee Law

We describe in more detail the process of dealing with decisions in the field of immigration and refugee law. All Canadian immigration decisions are retrieved from the Federal courts web site when they become public, and are then processed in order to produce two valuable pieces of information : the factsheet (See Figure 1) and an automatic summary of the decisions (Figure 2).

As the Court decisions in this domain are well structured, it is possible to identify three main parts and develop a specialized information extraction process for each:

Prologue a list of semi-structured information such as the docket number, the place and date of hearing, the judge's, plaintiffs' and defendants' names. Each piece of information is usually introduced by a specific label but the concept extraction and the determination of the matter of the decision require a more detailed analysis.

Decision a full-length text, structured in sections usually identified by titles or by specific sentences starting those sections. A typical decision is divided into six themes usually appearing in the following order: introduction, context, issues raised by the plaintiffs, reasoning, conclusion and the order. Some sections may be missing in some decisions, while additional sections may appear in other ones. The order in which sections appear may also vary.

Epilogue another list of semi-structured information such as the lawyers' and solicitors' names.

The information from the prologue and epilogue are kept in a database and an automatic table style summary is produced for the decision. The result is then reviewed by a lawyer from *NLP Technologies* who can make some manual adjustments. The overall result is reviewed by an editorial board before the information becomes available to the company's subscribers on the Web. This mix of automatic processing and manual review has been in operation for 4 years and has given very good results on Immigration decisions written in English. Using the parameterization process described below, we were able to extend, in the course of 2008, the system to decisions in the same field written in French and to decisions in tax and intellectual property laws. Two core ideas have presided to this re-engineering: the use of a linguistics aware technology and parameterization.

4   Linguistics–Aware Information Extraction Process

Canadian immigration decisions are available on the Web as HTML documents either in English or French depending on the language used at the hearing. A decision may naturally be relevant for Canadian lawyers no matter in which language it is written. Since HTML tags define the presentation of those decisions, rather than their structure, and since the presentation as well as its HTML definition is liable to evolve over time (and it has), we cannot rely on only these tags to identify the structure of the decisions. We thus analyze the text of the decision itself to discover the sentences of each section to appear in the summary.

Figure 6 shows a simplified view of the transformation pipeline combining different technologies to go from an original judgement as an HTML to an XML file that is saved in a data base from which the final summary, also in HTML, is

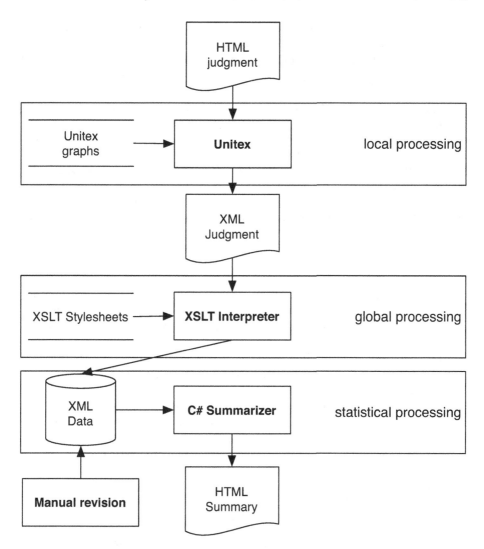

Fig. 6. System architecture going from the original to the summary. Unitex graphs are used for going from HTML to XML and for linguistic processing within a sentence or for short spans of text. XML Transformation Stylesheets (XSLT) are able to take into account long distance dependencies and the statistical computations for determining the most important sentences to appear in the summary are done by a C# program.

generated. *NLP Technologies* lawyers, through a specialized reviewer interface, can also change this XML file during the manual review process. This transformation process involves both local (within a sentence) processing, more global processing taking into account parts of the documents that can be farther apart and statistical processing for computing the salient sentences that will compose the final summary.

We decided to use technologies that are appropriate for each step of the transformation. Transductors allow a great flexibility in sentence processing, XSLT stylesheets are an efficient means for selecting and transforming longer spans of texts and a procedural language is used for computing the final statistics to select the final sentences that appear in the summary.

The unit of work in all transformation steps is the whole sentence in order to guarantee that the summary contains only original sentences that can be cited verbatim without having to consult the judgement. Transductors and stylesheets add *hidden* information to sentences of the original text to provide hints to the final statistical summarization module that decides for each sentence whether it will appear in the summary or not, and if so, in which thematic segment it will be put. Even the manual reviewers work at the level of sentence and choose to either add or remove a whole sentence or not; they are not allowed to modify the wordings of sentences.

4.1 Local Processing

A first step is thus to convert HTML documents into text files and then use linguistic cues to identify the decision structure as well as the relevant factual information. Fortunately, decisions follow a rather stereotypical pattern and use recurrent information identifiers or section headings. Such identifiers have several variants, but there are usually a fixed set of them.

We decided to use XML tags to identify text structure and relevant factual information, since there are several general-purpose XML-based processing tools, such as structure validation or document transformation tools. So our process will first eliminate most HTML tags and transform others into paragraph markers.

Relevant information will then be identified through linguistic cues, which are phrases identifiable through context-free grammars. As we are aiming for power and flexibility, we decided to make use of the transductor technology, namely Unitex[4], a descendant of INTEX [8], to identify, mark and transform spans of texts by means of regular expressions which provides the following advantages:

- Regular expressions are represented with graphs (see Figure 7 for an example) instead of complex sequences of operators and their base unit is the word rather than the character. Language-dependent character equivalences are appropriately handled.
- It works with a user-defined dictionary in which words and phrases may be assigned various user-defined syntactic or semantic categories which may in turn be used in graphs. Flexional categories and morphological criteria can be almost freely combined with those syntactic and semantic categories, enabling the expression of complex search criteria without ever having to translate those criteria into character patterns.
- Graphs may be used as subgraphs of other more complex graphs, enabling graph reusability.

---

[4] http://www-igm.univ-mlv.fr/~unitex/

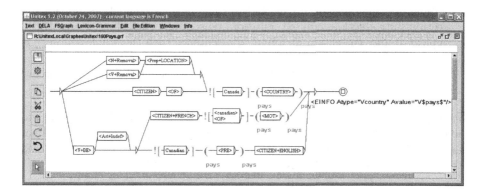

Fig. 7. A graph defines a set of paths matching words encountered in the text going from the entry node (the triangle on the far left) to the exit node (the circle containing a square) on the right. A node can match either be a single word (see **Canadian** above), or one word contained in a list defined in the dictionary (see **<COUNTRY>** above). When a path going from the entry to the exit has been found, information can be added (shown here in bold) to the original text. Here the occurrence detected is tagged with an XML tag named **EINFO** with attributes **ATYPE** having value country and **Avalue** having a value **pays** that was saved during matching this graph. This graph detects the country from which the applicant originates. The 4 paths out of the start state, from top to bottom, correspond respectively to: 1) a path that recognized phrases such as **his removal to Kenya**, 2) a path that recognizes phrases such as **[is scheduled to be] removed to Kenya**, 3) a path that recognizes phrases such as **[is a] citizen of Kenya**, 4) a path that recognizes phrases such as **[is a] Kenyan citizen** or **[est] citoyen kenyan**. Note that adjectives derived from country names, recognized by the last path, are not listed in the dictionary contrary to country names, which are listed.

- Parameterized graphs (explained in the next section) add even more flexibility to our processing.

Unitex graphs have the power and efficiency of regular expressions, with the additional benefits of linguistic awareness and much improved user-friendliness. These grammars recognize word patterns most often limited to a single sentence. Unitex processing of the judgements involve the use of 33 compiled graphs for transforming the HTML form of a judgement to a labeled XML file. An example of such a graph that detects the applicant's country of origin is displayed as Figure 7.

4.2 Global Processing

Although there is no theoretical limit on the span of input that can be processed by a Unitex transductor, in practice we have experienced many problems when the input is too long. Unitex is cumbersome for expressing long-range dependencies but there are however a few contextual or structural rules to implement, such as:

- A sentence that contains a pattern associated with salient phrases of a given section (introduction, context, citation, reasoning, conclusion). If a pattern, typical of a given section, is found in a sentence then the whole sentence is assigned to this section.
- All sentences of a paragraph following a sentence identified as a citation are also part of that citation.

We decided to express such structural rules with XSLT stylesheets applied to the resulting XML format of the documents. Using XML provides the additional benefit of checking the conformity of the document structure to the XML schema associated with decisions. The XSLT processing uses 10 templates.

4.3   Statistical Processing

To identify the sentences to appear in the summary, some statistical computations are involved such as the computation of TF·IDF scores and other numerical values. This process is done with a C# program that parses the XML document produced by the previous two steps. The HTML input files are about 30K characters long, corresponding to 16K words. On a stock desktop PC, the processing time for applying Unitex graphs, processing XSLT templates and computing statistics is about 40 seconds per judgement.

5   Parameterization of the Information Extraction Process

As partly shown in Figure 7, Unitex graphs can refer to words defined in a dictionary, a user-defined list of word forms associated with their root form as well as various syntactic and semantic categories and morphological features. It would be cumbersome to define all word forms by hand, especially in an inflected language like French in which semantic categories do not vary with the flexion. Unitex offers two types of dictionary definitions: the inflected dictionary, where it is possible to directly define word forms, and the non-inflected dictionary, which will be inflected by Unitex using an inflexion graph provided by the user. Such graphs are language dependent but are application domain independent.

Unitex offers an additional mechanism called the parameterized graph, which combines a generic graph containing variables and a parameter file. The latter is a text file containing the values to be taken by the variables. More precisely, each line of the parameter file will generate a subgraph, and the whole family of subgraphs will be integrated as a single graph. Each subgraph thus represents an alternative and the main graph is a disjunction of all those alternatives. In order to maximize the parameterization of our system, we have made an extensive use of the dictionary as well as of parameterized graphs, so that many graph updates can simply be made through the update of those parameter files followed by a graph recompilation. We have used Microsoft Excel to assemble the various parameter files and to simplify the data definition. Excel macros are used for validation and for cross-checks between those lists. Excel is also a user-friendly way of consulting, sorting and filtering those parameter lists.

Some operators such as X in-same-sentence-as Y or X near Y, not available in Unitex have been developed with auxiliary graphs, and can be used in those lists to implement complex rules: there is a fixed list of them however, since we did not want to implement a general rule compiler. In total, there are 10 worksheets in this Excel file: each of them parameterizes a specific aspect of the information extraction process. The dictionary itself contains 432 uninflected single words, 840 inflected single words or single words without any flexion and 812 phrases. Those figures combine both English and French entries. In a specialized information extraction setting like this one, we only have to deal with words that are used for segmenting the judgement or for identifying specific information like dates, names of parties. Most of the words encountered in the text are simply taken as is and will be given back verbatim if it happens that the sentence as a whole is chosen to appear in the summary.

6 Application to the Intellectual Property and Tax Law Domains

Once the information extraction process was completed for the immigration domain, a natural step was to extend it to other law domains of interest to *NLP Technologies*, namely the intellectual property and the tax law domains. In both cases, federal decisions were the only ones taken into consideration.

The intellectual property domain was very easy to integrate: decisions from this domain emanate from the same courts as the immigration ones, and thus follow the same structure. The main differences between both domains lie in the subjects, topics, and laws associated with one domain or the other, as well as to some specific dictionary entries. Since these data are parameterized in an Excel file, it was very easy to add French and English data relevant to the intellectual property domain to the file. Only very minor reorganizations of this file were required, such as adding a domain field to the topic and the subject worksheets in order to facilitate their maintenance. The integration of the intellectual property domain took about three weeks.

Integrating the tax law domain was more challenging: decisions from this domain can originate from the Federal Court or the Federal Court of Appeal, as is the case for the previous domains, but also from the Tax Court of Canada. Decisions from the latter differ in structure from those issued by the Federal Court or the Federal Court of Appeal, especially in the order and way in which the prologue and epilogue information is presented, and thus required not only an update of the Excel file, but also some modifications of the local processing step. Since we wanted to keep one single processing unit for all decisions, we just added a parameter that states whether a decision is issued form the Federal Court of the Federal Court of Appeal, on the one side, or from the Federal Tax Court on the other side. It must be emphasized that these differences were big enough to justify adding two Unitex graphs for the Federal Tax Court decisions

Table 1. Number of topics and subjects associated with each domain. The *All domain* (not shown) indicates topics and subjects independent of a specific domain. They are associated with court practice questions that arise in all domains.

Domain	Subjects		Topics	
	examples	nb	examples	nb
Immigration	Appeal by Permanent Resident, Appeal by Protected Person, Citizenship application, Entrepreneurs, Family class application, Inadmissibility, Investors, Convention Refugee Abroad class, Refugee Protection, Enforcement of Removal orders, Skilled workers, Stay of removal orders, Study Permit, Visitors, Work permit, Source country class ...	33	Assurances against torture, Child custody order, Deserters, Failure to seek protection, Gangs, Habeas Corpus, HIV-positive, Identity, Irreparable harm, Obligation to avoid risk of persecution, Oral interview, Religious conversion, Removal from record, Review of detention, Risk assessment, Street gangs, Vengeance, Visa officer ...	1864
Intellectual Property	Patent, Trademark, Copyright, Industrial Design ...	4	adding parties, deadwood, elastomer, ex turpi causa, processability, prodrugs, Pseudonyms, recording medium, representation by non-lawyer, titles, work product ...	745
Tax	Income Tax, Income Tax Québec, Unemployment/Employment Insurance, Excise Tax, Goods and Services Tax GST, Canada Pension Plan, Old Age Security, Petroleum and Gas, Cultural Property Export and Import, Customs, War Veterans, Softwood Lumber, Tax Court Practice, Aboriginals ...	14	acupuncture, automobile allowance, bill of costs, business investment losses, constructive trust, contents of appeal book, foreign-based documents, incarceration, investment brokers, lawyers' disbursements, motion to reconsider, unjust enrichment, vehicle fees ...	1880
All		1		46
Total		52		4535

and making two versions of two existing graphs, but those modifications are still very minor and did not call for a major rethinking of the whole processing chain.

A few more technical adaptations of the process were required, such as splitting the execution of the topic graphs into five steps rather than as a whole for performance reasons. Once again, despite those minor updates, the integration of the tax law domain took only about six weeks. Whatever the domain associated with a decision, the latter is processed in the same way. It must be noted that the local or global processing steps do not attempt to assign a domain to the decisions they process. Theoretically, an immigration topic or subject could thus be erroneously attributed to an intellectual property decision for instance. This is however very rare, and those mistakes can be corrected either by the statistical processing or the manual review that follow the local and global processing steps.

The integration of new domains (their number is indicated in Table 1) was thus almost effortless thanks to the parameterization approach described above to manage the immigration domain. This success in adding new domains shows that, although our methodology is primarily based on hand defined dictionaries and transductor graphs, these can be quickly adapted because we stay within the law domain for which the fundamentals stay constant.

7   Maintenance of the Information Extraction Process

The information extraction transductors were developed originally by the manual inspection of about 60 decisions in both English and French published in 2007. Only a few (about 5%) of current decision were not processed correctly and involved some manual adjustment either by correcting the formatting of the input or by adding new words to the dictionary.

We have also tested the transductors on 14 380 historical decisions published between 1997 and 2006. Only 15% of those decisions were incorrectly processed by the original information extraction process, i.e. the resulting XML document was not well-formed, usually because the beginning of a section was detected but not its end or vice-versa. This happens because these complementary elements are tagged independently. Resolving the problems caused by 9 decisions helped resolve the problems encountered in 49 additional decisions (over 90 decisions tested). In other words, a single problem occurred on average on 6.5 decisions among the 90 decisions on which corrections were tested. Among those 9 problems, 3 implied adding entries in the dictionary, 5 implied modifying existing graphs in order to improve their flexibility. We decided not to take any action on the last one which was caused by a misspelling in the decision. It is yet unclear whether our parameterization effort has been sufficient, since only 3 problems out of 8 could be solved without modifying any graph. We are just at the beginning of the correction process however, and we hope that, as time goes on, a higher proportion of problems will be solved through dictionary update, as well as we can hope that one single correction will have a positive impact on more decisions. Moreover, we know that decisions have been presented in a considerably more homogeneous way since 2003, so that historical results are worse than those obtained from current decisions.

Thus, we are confident that as time goes on, there will be increasingly less manual work to do by *NLP Technologies* legal staff, who will merely need to check that everything is all right for publication. This process is in production since the summer of 2008.

Farzindar [2] compared the approach underlying *DecisionExpress* and other state-of-the-art summarization systems, but we are not aware of any similar commercial legal summarization system.

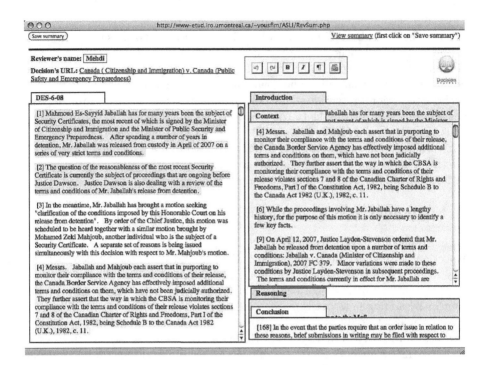

Fig. 8. Interface for the manual review (bottom left box of Figure 6) of the summaries produced by the automatic process within *DecisionExpress*. The left part shows all sentences of the judgement. Paragraphs selected by the automatic process are highlighted with a different color according to the theme they were assigned to. In this figure, the reviewer is working with the *Context* theme whose the tab is currently *opened*. To remove a paragraph from the summary, the reviewer double-clicks on a paragraph from the theme, to add a paragraph to this theme she double-clicks on a paragraph in the judgement. It is possible to add or remove single sentence but its content cannot be changed. This ensures that the original text is preserved in the summary. Simple sentence or paragraph highlighting (bold or italics) buttons are available at the top right. Once the reviewer is satisfied, the resulting summary can be saved in the database using the top left button.

Although we did not conduct any formal evaluation, the feedback given by the Federal Courts is that they find the results of the summaries produced and reviewed by *DecisionExpress* 100% precise and very useful. The electronic dissemination of the judicial decisions within the Federal Courts offices made possible by *DecisionExpress* also brought an interesting *environmental* benefit. The Federal Court used to print its weekly decisions for all of its judges about 1.5 million pages yearly. When the decisions were no longer used by the judges, they were picked up and stored. After the implementation of *DecisionExpress*, a poll taken amongst judges showed that a massive majority agreed that the Court should stop providing printed copies of the judgements.

Human reviewers find that about 70% of the sentences or paragraphs are identified correctly by the automatic system described in this paper. We are currently improving the system using statistical methods now that we have a corpus of reference summaries that *NLP Technologies* has produced over the years. Even though all summaries have to be validated by human reviewers, this process takes less than 15 minutes per decision. We expect the review process to be even faster now that we are implementing the specialized review process interface shown in Figure 8.

8   Conclusion and Perspectives

DecisionExpress is the first service in the world based on an automatic summarization system developed specifically for legal documents. It is implemented in a real-life environment and currently produces summaries for large collections of judgements (between 50 and 100 each week) written in English or French in the immigration domain.

In this article, we have presented our recent work with respect to extending the applicability of the system to French and to other domains such as tax and intellectual property law. The main idea was to elaborate on an information management platform to organize the linguistic cues and semantic rules to achieve a precise information extraction in different fields. The output of the system is systematically reviewed by a lawyer but the goal is to have the system do as much work as possible.

In order for a client to work in the language they are most comfortable with, the RALI and *NLP Technologies* have developed a bidirectional French and English statistical machine translation (SMT) engine for judgements [3]. The SMT output sentences are reviewed before publication, similar to the process used by *NLP Technologies* for summaries.

As the summaries are extracts of the original judgement, we are also developing an interface to keep track of revisions (removal of selected sentences by the system or adding of new sentences) done on the summaries so that the corresponding translated sentences now form the summary in the other official language of Canada.

NLP Technologies is currently studying the possibility of extending the system to other courts and countries. US courts are particularly targeted because of the number of decisions and the proximity to Canada, but they are quite challenging because of different source formats and a different legal system.

Acknowledgments

We thank the CRIM-Precarn Alliance program and National Research Council Canada - Industrial Research Assistance Program (NRC-IRAP) for partially funding this work. We acknowledge the collaboration from the Federal Courts and their feedback. We sincerely thank our lawyers Pia Zambelli and Diane Doray. The authors also thank Fabrizio Gotti, Mehdi Yousfi-Monod, Farzaneh Kazemi and Jimmy Collin for technical support and Elliott Macklovitch for many fruitful discussions.

References

1. Farzindar, A., Lapalme, G.: LetSUM, an automatic Legal Text Summarizing System. In: Gordon, T.F. (ed.) Legal Knowledge and Information Systems. Jurix 2004: the Sevententh Annual Conference, pp. 11–18. IOS Press, Berlin (2004)
2. Farzindar, A.: Résumé automatique de textes juridiques. Ph.D. Thesis, Université de Montréal and Université de Paris IV-Sorbonne (2005)
3. Gotti, F., Farzindar, A., Lapalme, G., Macklovitch, E.: Automatic translation of court judgements. In: Proceedings of the Eighth Conference of the Association for Machine Translation in the Americas, Waikiki, Hawai, pp. 370–379 (2008)
4. Grover, C., Hachey, B., Korycinski, C.: Summarising legal texts: Sentential tense and argumentative roles. In: Radev, D., Teufel, S. (eds.) HLT-NAACL 2003 Workshop: Text Summarization (DUC 2003), Edmonton, Alberta, Canada, pp. 33–40 (2003)
5. Mailhot, L.: Decisions: a handbook for judicial writing, Editions Yvon Blais, Québec, Canada (1998)
6. Moens, M.F.: Summarizing court decisions. Information Processing and Management 43, 1748–1764 (2007)
7. Moens, M.F., Uyttendaele, C., Dumortier, J.: Abstracting of legal cases: the potential of clustering based on the selection of representative objects. Journal of the American Society for Information Science 50(2), 151–161 (1999)
8. Silberztein, M.D.: Dictionnaires électroniques et analyse automatique de textes. Le système INTEX, Paris, Masson (1993)
9. Smith, J.C., Deedman, C.: The application of expert systems technology to case-based law. In: Proceedings of ICAIL 1987, pp. 84–93 (1987)

Evaluation Metrics for Consistent Translation of Japanese Legal Sentences

Yasuhiro Ogawa, Kazuhiro Imai, and Katsuhiko Toyama

Graduate School of Information Science, Nagoya University Furo-cho, Chikusa-ku,
Nagoya, 464-8603, Japan
yasuhiro@is.nagoya-u.ac.jp

Abstract. We propose new translation evaluation metrics for legal sentences. Since most previous metrics, that have been proposed to evaluate machine translation systems, prepare human reference translations and assume that several correct translations exist for one source sentence. However, readers usually believe that different translations denote different meanings, so that the existence of several translations of one legal expression may confuse them. Therefore, since translation variety is unacceptable and consistency is crucial in legal translation, we propose two metrics to evaluate the consistency of legal translations and illustrate their performances by comparing them with other metrics.

Keywords: CIEL, Translation Evaluation Metric, Legal Translation, Consistency, BLEU.

1 Introduction

Recently, the social demand for the translation of Japanese statutes into foreign languages has been increasing in order to, for instance, conduct international transactions more smoothly, promote international investment in Japan, or support legal reform in developing countries. Since Japanese statutes have been individually translated by government ministries or private publishing companies, translation equivalents have been inconsistent among translated documents. For example, the legal term "弁護士(*bengoshi*)" has been translated as "*attorney*," "*barrister*," and "*lawyer*," which have different meanings in English. Therefore if "*attorney*" is used in one document, while "*lawyer*" is used in another document for the same term "弁護士," it is hard to recognize that both English words denote the same word in the source documents, or, in some cases, they may confuse readers. For this reason, the same translation equivalent should be used for the same term: consistent translation is required.

To solve this problem, the Japanese government has compiled a *Japanese-English Standard Bilingual Dictionary*[1] [3] [9] for legal technical terms occurring in Japanese statutes, which includes about 3,700 Japanese entries and about 4,700 English equivalents. Now, Japanese statutes are being translated in compliance with this dictionary by the government. Then, the next task is quality evaluation of the translations, which should be done also in compliance with the dictionary.

---

[1] http://www.japaneselawtranslation.go.jp/

E. Francesconi et al. (Eds.): Semantic Processing of Legal Texts, LNAI 6036, pp. 235–248, 2010.
© Springer-Verlag Berlin Heidelberg 2010

However, since one term sometimes has several translation equivalents, a suitable one in context should be used in a translation. For example, in the Standard Bilingual Dictionary, the term "免除する(menjo-suru)" has six equivalents: "release," "exempt," "waive," "exculpate," "remit," and "immunize." We should choose the most suitable one among them depending on the context. Although notices for the choice might be roughly given to some equivalents in the dictionary, registering every detailed criterion for the choice in the dictionary is difficult. Thus it is insufficient to only rely on the dictionary for consistent translations.

Therefore we need an automatic evaluation metric for consistent translations. Several translation metrics have been proposed: BLEU [8], Word Error Rate (WER), Position independent Word Error Rate (PER) [6], METEOR [1] and NIST [2]. However, these metrics were designed to evaluate machine translation systems and do not evaluate human translations. Since their basic ideas are to compare machine translations with human reference translations that are considered correct, they require such reference translations.

On the other hand, we cannot prepare human reference translations for the evaluation of legal translations. In fact, if we can prepare *correct* reference translations, we no longer need other translations. Therefore, we must prepare alternative references.

In this study, we focus on the fact that most Japanese legal sentences are described in terms of fixed expressions. This is because the Cabinet Legislation Bureau reviews most Japanese statutes and controls the use of legal terms and expressions in the statutes during the process of drafting. From the viewpoint of consistency, the same fixed expression should have the same translation. For this purpose, we used a legal parallel corpus and compared translations with it.

In particular, for the corpus, we used the translations of Japanese statutes released by the Japanese government[1] [3], including 17,793 Japanese legal sentence types. We use this parallel corpus instead of human reference translations. However, the translations in the corpus may not be translations of source sentences but translations of similar sentences to the sources. We call such translations *pseudo reference translations* (PRTs).

We tried to use the BLEU metric [8] to compare a candidate translation with PRTs. However, the BLEU metric is not convenient for our comparisons, since PRTs are not translations of source sentences. To solve this problem, we modified the BLEU metric tailored with PRTs and we named it **CIEL**.

We applied the CIEL metric to three kinds of translations of the Labor Standard Act: by the Japanese government, a publishing company, and the Google translation tool. As a result, the CIEL metric succeeded in distinguishing the translations by the government from those by the publishing company, but the BLEU metric could not. In addition, we compared the CIEL metric with other evaluation metrics and showed its effectiveness.

This paper is organized as follows: in Section 2, we introduce the BLEU metric as the baseline. Next, we propose our evaluation metrics in Section 3. Then we describe some evaluation experiments in Section 4. Finally, Section 5 is a conclusion.

2   BLEU

The BLEU metric [8] is an automatic evaluation metric for machine translation. Its basic idea compares n-grams occurring in the candidate translation, which is a machine translation sentence for a given source sentence, with n-grams occurring in the human reference translations. Since several translations are possible for one source sentence, the BLEU metric prepares multiple human translations as references. For comparison, the following precision score p_n is calculated:

$$p_n = \frac{\sum\limits_{S \in TranslationDocument} \sum\limits_{n\text{-}gram \in S} Count_{clip}(n\text{-}gram)}{\sum\limits_{S \in TranslationDocument} \sum\limits_{n\text{-}gram \in S} Count(n\text{-}gram)} , \quad (1)$$

where $Count(n\text{-}gram)$ is the number of occurrences of $n\text{-}gram$ in the candidate translation S. $Count_{clip}(n\text{-}gram)$ is also the number of occurrence of $n\text{-}gram$ in S, but if it is greater than the maximum number of occurrences of $n\text{-}gram$ that occurs in any single reference translation, $Count_{clip}(n\text{-}gram)$ is equal to the maximum number. Notice that if $n\text{-}gram$ does not occur in any reference translations, $Count_{clip}(n\text{-}gram)$ is 0. The external sum ranges over all candidate translations in the document, that is, the BLEU metric evaluates entire translation documents.

Next, if the candidate translation is shorter than its reference translations, the denominator of the above formula becomes smaller so that p_n becomes larger. To penalize this situation, the BLEU metric computes brevity penalty (BP):

$$BP = \begin{cases} 1 & \text{if } c > r \\ e^{1-r/c} & \text{if } c \leq r \end{cases}, \quad (2)$$

where c is the length of the candidate translation and r is its effective reference length.

Finally, introducing positive weights w_n based on the value of n, the final BLEU score is defined as follows:

$$BLEU = BP \cdot \exp\left(\sum_{n=1}^{N} w_n \log p_n\right). \quad (3)$$

Usually, the upper of n is set to be $N = 4$ and uniform weights $w_n = 1/N$. Using from unigrams to 4-grams, the BLEU metric evaluates both adequacy and fluency of candidate translations, where the adequacy indicates how much information is retained in the translation and the fluency indicates to what extent the translation reads like good English.

3   Evaluation Metric Considering Consistency

The BLEU metric needs several human reference translations that are considered *correct*. For the evaluation of translation of legal sentences, however, we cannot prepare reference translations. In fact, if we have a *correct* translation of a legal sentence, we do not need to evaluate other translations any more. So we have to evaluate human translations without *correct* references.

Here, notice that Japanese legal sentences have many fixed expressions. For example, the sentences that provide effective dates of each act have the following expressions:

Source 1: この法律は、会社法の施行の日から施行する。
Source 2: この法律は、行政手続法の施行の日から施行する。

For consistent translation, such fixed expressions as "この法律は、...の施行の日 から施行する。," shown as underlined, should be translated into identical expressions. Therefore Sources 1 and 2 should be respectively translated as follows:

Translation 1: This Act shall come into force as from the date of enforcement of the Companies Act.
Translation 2: This Act shall come into force as from the date of enforcement of the Administrative Procedure Act.

We used a parallel corpus of Japanese statutes to evaluate such consistency. First, we retrieved similar sentences to a given source sentence and collect their translations, which can be considered reference translations. However, since such translations may not be translations of source sentences, we call them *pseudo reference translations* (PRTs). To compile PRTs, we used legal translations released by the Japanese government[1] [3]. We assume they are suitable translations since they were made in compliance with the Standard Bilingual Dictionary to improve consistency and reliability of the translations. Thus, the translations in the PRTs can be considered adequate and fluent in terms of consistency.

We describe the details of the compilation of PRTs in Section 3.1 and how to evaluate the translations in Sections 3.2 and 3.3, respectively.

3.1   Acquisition of Pseudo Reference Translations

We used a hierarchical clustering method [4] to obtain a set of PRTs. We divided the source sentences in the corpus into clusters and selected the closest one to a given source sentence. Since such clusters contain similar sentences to the source, we collected their translations as PRTs. The following shows the details of the clustering method.

First, we identified a subset of source sentences, since the cost of clustering tasks for all sentences is considered to be too high. Here we used the peculiarity of Japanese language, that is, main predicates occur at the end of sentences and play an important role in sentences. So we split the source sentences by their last morphemes and reduced the clustering cost.

Next, we deleted all *bunsetsus*,[2] except the last one, those depending on the last one, and those depending on them. This is to delete non-fixed expressions from the sentences.

For example, the following *bunsetsus* are left after the deletion in Source 1 consisting of six *bunsetsus*:

---

[2] A *bunsetsu* is a linguistic unit in a Japanese sentence and roughly corresponds to a basic phrase in English.

この　法律は、　会社法の　施行の　日から　施行する。

Fig. 1. *Bunsetsu* deletion

1. "施行する (*shall come into force*)";
   this is the last *bunsetsu*.
2. "法律は (*Act*)" and "日から (*as from the date*)";
   these depend on the last *bunsetsu* "施行する."
3. "この (*This*)" and "施行の (*of enforcement*)";
   "この" depends on "法律は" and "施行の" depends on "日から."

This result is illustrated in Fig. 1. Source 1 becomes

この法律は、施行の日から施行する。
(*This Act shall come into force as from the date of enforcement*).

In order to analyze dependency relations between *bunsetsus*, we used CaboCha[5], which is a Japanese dependency/syntactic parser based on machine learning and achieves about 90% accuracy.

After transforming the source sentences as above, we applied hierarchical clustering. We used the group average method and the morpheme-based edit distance. The distance between two sentences is defined as the minimum number of operations needed to transform one sentence into the other, where an operation is the one of the insertion, deletion, or substitution of a single morpheme. However this distance is sensitive to the sentence length, so we normalized it into interval $[0, 1]$ by dividing by the sentence length.

We used the resulting clusters as PRTs except those containing only one sentence since such clusters are unreliable for evaluation.

Furthermore, we noticed that fixed sentences are used in many statutes. For example, the sentence

この法律は、公布の日から施行する。
(*This Act shall come into force as from the day of promulgation.*)

appears in many statutes. From the viewpoint of consistent translation, the same source sentences should be translated into the same translation. If the same sentence is included more than once in the corpus, we use the translations of the sentence as reference translations instead of the cluster.

3.2   Modifying BLEU Metric

Since the BLEU metric considers both adequacy and fluency of the translation, we might easily consider it an evaluation metric for our purpose. However, there are some problems when using PRTs. Thus we modified the metric as described in the following subsections.

Problems with BLEU Metric.   Section 3.1 showed how to get PRTs. Although the reference translations used in the BLEU metric are the ones of a given source sentence, *pseudo* reference translations are not. This causes some evaluation problems. For example, consider the following two candidate translations:

Source: 次に掲げる者は、委員となることができない。
Candidate 1: The following persons may not act as <u>committee members</u>:
Candidate 2: The following persons may not act as <u>members</u>:

For comparison, we prepared the following two PRTs:

Pseudo Reference 1: The following persons may not act as directors:
　　(次に掲げる者は、取締役となることができない。)
Pseudo Reference 2: The following persons may not act as accounting auditors:
　　(次に掲げる者は、会計監査人となることができない。)

Both Candidates 1 and 2 obviously resemble each other, and the only difference is the equivalent of "委員": "*committee members*" and "*members*." Both translations have identical adequacy, even though their BLEU scores are different. Particularly, p_1, shown in Section 2, is calculated by dividing the number of unigrams occurring in both a candidate and any its references by the number of unigrams occurring in the candidate. Therefore, p_1 of Candidate 1 is 7/9 since it contains nine unigrams and seven occur in the references. In the same way, p_1 of Candidate 2 is 7/8. The length of the equivalents of "委員" affects the scores, but it is not desirable since they do not occur in any references. In other words, n-grams occurring in the candidate but not in the PRTs may reduce the BLEU scores too much.

Introducing Weight.   Since PRTs may not be translations of the source sentence, some n-grams occurring in the candidate translation may not occur in the PRTs, reducing the BLEU score, as shown in the above example.

This suggests that we should consider only the n-grams that occur in the PRTs. In addition, we assume that if an n-gram occurs in many PRTs, it may occur in the candidate translation. Therefore we introduce a weight $w(n\text{-}gram)$ that indicates the ratio of sentences containing n-*gram* and propose the following weighted BLEU metric (**BLEU-W**):

$$w(n\text{-}gram) = \frac{\text{\# of sentences with } n\text{-}gram \text{ in PRTs}}{\text{\# of sentences in PRTs}}, \tag{4}$$

$$p_n = \frac{\sum_{n\text{-}gram \in S} Count_{clip}(n\text{-}gram) * w(n\text{-}gram)}{\sum_{n\text{-}gram \in S} Count(n\text{-}gram) * w(n\text{-}gram)}, \tag{5}$$

$$\text{BLEU-W} = \exp\left(\sum_{n=1}^{N} w_n \log p_n\right),\tag{6}$$

where the multiplication of BP as the BLEU metric is not included in the BLEU-W metric. BP is used as a penalty for shorter candidate translations. However, since PRTs are not translations of source sentences, the lengths of PRTs have nothing to do with those of candidate translations. In fact, BP causes a negative effect in preliminary experiments so that we delete BP from the BLEU-W metric.

The original BLEU metric evaluates the whole of translation documents, while the BLEU-W metric evaluates each sentence. By this modification, $w(n\text{-}gram)$ would be 0 if $n\text{-}gram$ does not occur in the PRTs since it does not contribute to the score.

Problems with BLEU-W Metric. Although the above BLEU-W metric succeeded in removing the negative effects of n-grams that occur only in the candidates, it causes another problem. Consider the following candidates to the source in Section 3.2

Candidate 2: The following persons <u>may</u> not act as members:
Candidate 3: The following persons <u>can</u> not act as members:

The only difference between them is "*may*" and "*can*." Since "*may*" is used in both Pseudo References 1 and 2 while "*can*" is not in them, Candidate 2 is more appropriate than Candidate 3. Despite this, the BLEU-W metric does not consider "*can*" because it does not occur in the PRTs, i.e., $w(\text{"}can\text{"}) = 0$. Thus p_1 of Candidate 2 is 7/7 and 6/6 for Candidate 3; they cannot be distinguished.

Considering Recall. To solve this problem, we should use more n-grams for the score calculation. In the BLEU-W metric, we assumed that an n-gram occurring in many PRTs *may* occur in the candidate translation and we did not consider the n-grams that do not occur in the candidate.

However, from the viewpoint of consistency, an n-gram occurring in many PRTs *should* occur in the candidate translation since the PRTs are the translations of sentences similar to the source sentence, and such n-grams may be fixed expressions.

Therefore we define $TopRef(n, \alpha)$ as the set of n-grams occurring more than the ratio $\alpha(0 \le \alpha \le 1)$ in the PRTs as follows:

$$TopRef(n, \alpha) = \{n\text{-}gram \in Ref \,|\, w(n\text{-}gram) > \alpha\},\tag{7}$$

where Ref is the set of n-grams in the PRTs. Using $TopRef(n, \alpha)$, we modify p_n as follows:

$$p_n = \frac{\displaystyle\sum_{n\text{-}gram \in S \cup TopRef(n,\alpha)} Count_{clip}(n\text{-}gram) * w(n\text{-}gram)}{\displaystyle\sum_{n\text{-}gram \in S \cup TopRef(n,\alpha)} \max(Count(n\text{-}gram), 1) * w(n\text{-}gram)},\tag{8}$$

$$\text{CIEL} = \exp\left(\sum_{n=1}^{N} w_n \log p_n\right).\tag{9}$$

We call this metric **CIEL** (ConsIstency Evaluation for Legal documents). This metric can be considered a modification of BLEU with a recall-oriented strategy that came from the ROUGE metric [7].

3.3   Compliance Rate

Since PRTs may not be translations of source sentences, the BLEU-W and the CIEL metrics can evaluate only expressions occurring in the PRTs. For example, both metrics cannot evaluate the equivalents of "委員" in the Candidates 1 and 2 in Section 3.2 since the equivalents do not occur in the PRTs. Therefore both BLEU-W and CIEL are insufficient metrics for adequacy.

However, we can evaluate translation adequacy, i.e., whether adequate equivalents were used, using the Standard Bilingual Dictionary (SBD) which is mentioned in Section 1. The adequacy can be evaluated by considering standard equivalents in candidate translations.

Thus we define compliance rate (CR) to evaluate adequacy:

$$CR = \frac{\sum\limits_{Source\ sentences} \#\ of\ occurrences\ of\ SBD\ equivalents}{\sum\limits_{Source\ sentences} \#\ of\ occurrences\ of\ SBD\ entries} . \tag{10}$$

4   Evaluation Experiment

We evaluated the proposed metrics by comparing them with other metrics through experiments.

4.1   Experimental Targets

We evaluated the BLEU-W and the CIEL metrics by calculating the scores for the translations of the Labor Standards Act, which contains 242 sentences. We prepared three kinds of translations: by the Japanese government, a publishing company, and a Google translation tool. We calculated BLEU without BP,[3] BLEU-W, and CIEL scores for each translation as well as CR.

The government translation was done by legal specialists using the SBD[1] [3]. The company translation was also done by legal specialists without using the SBD. Since the government translation is based on the SBD, it is expected more consistent than the company one. The Google translation was the result of the machine translation system created by Google[4], so that it can be considered not better than the others. Therefore the proposed metric is expected to highly score them in this order.

For the compilation of PRTs, we used the parallel corpus of Japanese statutes translated by the Japanese government[1], which include 17,793 Japanese sentence types of 84 acts and bylaws, excluding the Labor Standards Act and their translations as 20,154 English sentence types; one Japanese sentence sometimes has several English translated sentences.

In the CIEL metric, we set parameter α of $TopRef$ to 0.5.

---

[3] BP has a negative effect with PRTs as mentioned in Section 3.2.
[4] http://www.google.com/translate_t

Table 1. Scores of BLEU, BLEU-W, and CIEL

Distance to	BLEU			BLEU-W			CIEL		
the cluster	gov.	company	Google	gov.	company	Google	gov.	company	Google
$0.0 < d \leq 0.1$	0.386	0.365	0.158	0.964	0.980	0.971	0.682	0.607	0.346
$0.1 < d \leq 0.2$	0.340	0.343	0.153	0.949	0.956	0.958	0.731	0.708	0.386
$0.2 < d \leq 0.3$	0.326	0.327	0.168	0.986	0.981	0.979	0.619	0.592	0.413
$0.3 < d \leq 0.4$	0.177	0.170	0.104	0.966	0.965	0.978	0.525	0.498	0.353
$0.4 < d \leq 0.5$	0.168	0.153	0.119	0.972	0.973	0.957	0.423	0.393	0.328
$0.5 < d \leq 0.6$	0.109	0.124	0.097	0.947	0.953	0.949	0.334	0.330	0.275
$0.6 < d \leq 0.7$	0.087	0.083	0.075	0.917	0.914	0.910	0.333	0.326	0.293
$0.7 < d \leq 0.8$	0.095	0.095	0.089	0.907	0.901	0.953	0.283	0.261	0.241
$0.8 < d \leq 0.9$	0.071	0.074	0.088	0.958	0.961	0.941	0.118	0.115	0.109
$0.9 < d \leq 1.0$	0.156	0.143	0.131	1.000	1.000	1.000	0.293	0.302	0.283
Average	0.157	0.157	0.106	0.949	0.951	0.951	0.413	0.396	0.305

4.2 Experimental Results

Before calculating the scores for the candidate translations, we divided the 17,793 Japanese sentence types into 1,910 clusters and selected the closest cluster to each source sentence as mentioned in Section 3.1. But we could not calculate the score for 22 of the 242 sentences in the Labor Standards Act for the following reasons. In eight sentences, the closest cluster could not be selected since the last morphemes did not occur in any cluster. Eight other sentences had no closest cluster, since their distances to any clusters were equal to 1, that is, there were no clusters similar to them. In the remaining six sentences, since their closest clusters contained only one sentence, they were unreliable.

Thus we calculated the BLEU, BLEU-W, and CIEL scores for the translations of the remaining 220 sentences, and the result is shown in Table 1.

As seen in the average values in Table 1, the BLEU-W metric calculated similar scores for all the three kinds of translations, that is, it could not determine which translation was better. However, the BLEU metric showed significant differences between the Google translation and the others, while not between the government translation and the company one. On the contrary, the CIEL metric showed a significant difference between the government translation and the company one as well as between the Google one and the others.

When the distance between the source sentence and its closest cluster is small, the CIEL metric has a large difference between the government and the company. Table 2 shows the scores divided by each of the Google ones so that it makes clear the difference. From this, we conclude that the CIEL metric is reliable when the distance is small. However, the CIEL metric is unreliable when the distance is large, and we further discuss this in the next subsection.

On the other hand, the CR score, shown in Table 3, has a desirable order.

Table 2. Proportional scores of BLEU, BLEU-W, and CIEL

Distance to the cluster	BLEU			BLEU-W			CIEL		
	gov.	company	Google	gov.	company	Google	gov.	company	Google
$0.0 < d \leq 0.1$	2.44	2.31	1.00	0.99	1.01	1.00	1.97	1.76	1.00
$0.1 < d \leq 0.2$	2.22	2.24	1.00	0.99	1.00	1.00	1.90	1.83	1.00
$0.2 < d \leq 0.3$	1.94	1.95	1.00	1.01	1.00	1.00	1.50	1.43	1.00
$0.3 < d \leq 0.4$	1.71	1.64	1.00	0.99	0.99	1.00	1.49	1.41	1.00
$0.4 < d \leq 0.5$	1.40	1.28	1.00	1.02	1.02	1.00	1.29	1.20	1.00
$0.5 < d \leq 0.6$	1.12	1.28	1.00	1.00	1.00	1.00	1.21	1.20	1.00
$0.6 < d \leq 0.7$	1.15	1.10	1.00	1.01	1.00	1.00	1.14	1.11	1.00
$0.7 < d \leq 0.8$	1.07	1.07	1.00	0.95	0.95	1.00	1.17	1.08	1.00
$0.8 < d \leq 0.9$	0.80	0.84	1.00	1.02	1.02	1.00	1.08	1.05	1.00
$0.9 < d \leq 1.0$	1.19	1.09	1.00	1.00	1.00	1.00	1.04	1.07	1.00
Average	1.48	1.48	1.00	1.00	1.00	1.00	1.35	1.30	1.00

Table 3. Compliance rate

Translator	# of entries whose equivalents occur in the translation	CR
government	2,042	0.779
company	1,765	0.674
Google	1,533	0.585

(# of entries occurring in Labor Standards Act: 2,620)

4.3 Discussion

In the average score of the CIEL metric shown in Table 1, the government translation outperformed the company one. However, examination of each sentence reveals that some have undesirable results, meaning the company translation has a higher score than the government one. This is caused by the following two reasons.

First, when the distance between the source sentence and its closest cluster is small, the company translation outperforms the government one against our assumption. Fig. 2 shows such a case, where the closest cluster of the source sentence

"前項の委員会は、次の各号に適合するものでなければならない。"

has sentences that include the following fixed expression:

"...の...は、次の各号に適合するものでなければならない。".

Source Sentence: 前項の委員会は、次の各号に適合するものでなければならない。

Government Translation: The committee set forth in the preceding paragraph <u>must</u> conform to the following items:

Company Translation: The committee mentioned in the preceding paragraph <u>shall</u> conform to each of the following items:

Pseudo Reference Translation 1: The statement of the detailed explanation of the invention as provided in item 3 of the preceding Paragraph <u>shall</u> comply <u>with each of the following items:</u>
(前項第三号の発明の詳細な説明の記載は、次の各号に適合するものでなければならない。)

Pseudo Reference Translation 2: The statement of the scope of claims as provided in paragraph 2 <u>shall</u> comply <u>with each of the following items:</u>
(第二項の実用新案登録請求の範囲の記載は、次の各号に適合するものでなければ ならない。)

Pseudo Reference Translation 3: The statement of the scope of claims as provided in paragraph 2 <u>shall</u> comply <u>with each of the following items:</u>
(第二項の特許請求の範囲の記載は、次の各号に適合するものでなければならない。)

Fig. 2. The case where the company translation outperforms the government one

The government translation uses an equivalent "*must*" for "なければならない(*nakereba-naranai*)." However, the PRTs suggest that it should be translated not as "*must*" but "*shall*." Since the company translation uses "*shall*" in the PRTs, its CIEL outscores the government. In the same way, while "次の各号に(*tsugino-kakugouni*)" should be translated into "*with each of the following items*," the government translation uses "*to the following items*," reducing its CIEL score. As you see, the company translation is better than the government one in this case. The CIEL metric gives a higher score to the better translation, so that this metric is desirable.

Second, when the distance between the source sentence and its closest cluster is not small, the CIEL metric is unreliable since the source sentence does not resemble the sentences in the cluster. In such a case, the n-gram that should be used in the candidate translation does not occur in the PRTs and the n-gram contained in $TopRef(n, \alpha)$ does not occur in the candidate translation, either. As a result, the CIEL scores become small and the difference between the government and the company translation scores also becomes small, as shown in Table 1. So the company translation score sometimes becomes bigger than the government one. Therefore the CIEL metric cannot determine which is better if a source sentence has a large distance to its closest cluster. To solve this problem, we need to collect a more reliable parallel corpus and make a cluster closer to a source sentence.

4.4 Comparison with Other Evaluation Metrics

To compare the CIEL metric with other evaluation metrics, we calculated 95% confidence intervals based on different samplings of the test data. This comparison is proposed for two machine translation systems [10].

First, create test suites $T_0, T_1, \ldots, T_B$, where T_1 to T_B are artificial test suites created by resampling T_0. Then, system X scored x_0 on T_0 and system Y scored y_0. The discrepancy between systems X and Y is $\delta_0 = x_0 - y_0$. Repeat this process on every B test suite and we have B discrepancy scores: $\delta_1, \delta_2, \ldots, \delta_B$. From these B discrepancy scores, find the middle 95% of the scores (i.e. the 2.5^{th} percentile: $score_{low}$ and the 97.5^{th} percentile $score_{up}$). $[score_{low}, score_{up}]$ is the 95% confidence interval for the discrepancy between machine translation systems X and Y. If the confidence interval does not overlap with zero, the difference between systems X and Y is statistically significant.

In our comparison, neither are machine translations, but X is the government translation of the Labor Standards Act, and Y is the company one. If an evaluation metric can claim a significant difference between the two translations, the metric is desirable.

We compared CIEL with Word Error Rate (WER), Position independent Word Error Rate (PER) [6], METEOR [1], NIST [2], ROUGE-N [7] and BLEU.

The Labor Standards Act contains 242 sentences. However, from the above experiment, when the distance between a source sentence and its closest cluster exceeded 0.5, such a cluster is not reliable as PRTs. So we only used 94 sentences that have a closer cluster than a distance of 0.5. We also set $B = 2000$ as the number of times to repeat the process. Table 4 shows the results.

Since the confidence interval does not overlap with zero at PER, ROUGE-1, and CIEL, only these three metrics can claim that the government and the company translations are significantly different. We also count the number of undesirable cases, which are the scores when the government translation is less than the company one, as shown in the most right column of Table 4. The CIEL metric has the fewest cases, so it is the most desirable metric. Notice that the PER metric has few undesirable cases. However it only considers words and not word order; it only evaluates adequacy. Therefore the CIEL metric is the most desirable because it evaluates both adequacy and fluency.

Table 4. Confidence intervals for discrepancy between two translations

Metric	$Score_{low}$	$Score_{up}$	# of undesirable cases
$1 - WER^{\dagger}$	-0.014	0.017	795
$1 - PER^{\dagger}$	0.004	0.022	4
METEOR	-0.004	0.020	162
NIST	-0.150	0.217	704
ROUGE-1	0.004	0.035	17
ROUGE-2	-0.001	0.022	92
BLEU	-0.006	0.022	259
BLEU-W	-0.009	0.004	1451
CIEL	**0.011**	**0.052**	**2**

† We used the values subtracted from 1 for WEP and PER since they score lower if the translation is better, while other metrics score higher.

5 Conclusion

In this paper, we proposed two consistency evaluation metrics for legal translations: the CIEL metric with pseudo reference translations and compliance rate CR with the Standard Bilingual Dictionary. The CIEL metric is based on n-gram alignment scoring and clustering algorithms, both of which are suitable for Japanese legal documents that contain recurrent phrasal structures. In particular, the reason why we can use pseudo reference translations is that most Japanese legal sentences are described in terms of fixed expressions.

We also confirmed that these metrics can evaluate several translations of one source sentence from the viewpoint of consistency.

Since the CIEL metric requires suitable pseudo reference translations, collecting consistent legal translations for them is future work. In addition, the CIEL metric is relative, that is, it determines which is better when several candidate translations are given. We also need an absolute metric that can determine whether one candidate translation is consistent. Furthermore, we want to examine how the CIEL metric correlates to the intuitive evaluation by human experts.

We intend to use the proposed metrics in the English Translation Project of Japanese Statutes[1] to determine whether the first translation of a statute is appropriate for the project database that aims for consistency and reliability of the translation.

Acknowledgements. The authors would like to thank Professor KASHIWAGI Noboru, Chuo Law School, Chair of Experts Council for Translation of Japanese Laws and Regulations into Foreign Languages, Ms. SHIMADA Chiko, Deputy Director, Cabinet Secretariat, and Mr. OHYA Tai, Prosecutor, Ministry of Justice, for their cooperation. Thanks are also owed to Professor MATSUURA Yoshiharu, Associate Professors KAKUTA Tokuyasu and Frank BENNETT, and Mr. SANO Tomoya, Graduate School of Law, Nagoya University, for their discussions and suggestions. This research project is partly supported by a Grant-in-Aid for Scientific Research (B) from the Japan Society for the Promotion of Science.

References

1. Banerjee, S., Lavie, A.: Meteor: An automatic metric for mt evaluation with improved correlation with human judgements. In: Proceedings of the ACL Workshop on Intrinsic and Extrinsic Evaluation Measures for Machine Translation and/or Summarization, pp. 65–72 (2005)
2. Doddington, G.: Automatic evaluation of machine translation quality using n-gram co-occurrence statistics. In: Proceedings of the second international conference on Human Language Technology Research, pp. 138–145 (2002)
3. Study Council for Promoting Translation of Japanese Laws and Regulations into Foreign Languages: Final report (2006), http://www.cas.go.jp/jp/seisaku/hourei/report.pdf
4. Jain, A.K., Murty, M.N., Flynn, P.J.: Data clustering: a review. ACM Computing Surveys 31(3), 264–323 (1999)

5. Kudo, T., Matsumoto, Y.: Japanese Dependency Analysis using Cascaded Chunking. In: Proceedings of the International Conference on Computational Linguistics, pp. 1–7 (2002)
6. Leusch, G., Ueffing, N., Ney, H.: A novel string-to-string distance measure with applications to machine translation evaluation. In: Proceedings MT Summit IX, pp. 240–247 (2003)
7. Lin, C.: Rouge: a package for automatic evaluation of summaries. In: Proceedings of Text Summarization Branches Out, Workshop at the ACL 2004, pp. 74–81 (2004)
8. Papineni, K., Roukos, S., Ward, T., Zhu, W.: Bleu: a method for automatic evaluation of machine translation. In: Proceedings of the 40th Annual Meeting on Association for Computational Linguistics, pp. 311–318 (2002)
9. Toyama, K., Ogawa, Y., Imai, K., Matsuura, Y.: Application of word alignment for supporting translation of Japanese statutes into English. In: van Engers, T.M. (ed.) Legal Knowledge and Information Systems JURIX 2006: The Nineteenth Annual Conference, pp. 141–150 (2006)
10. Zhang, Y., Vogel, S.: Measuring confidence intervals for the machine translation evaluation metrics. In: Proceedings of the Tenth Conference on Theoretical and Methodological Issues in Machine Translation (TMI), pp. 85–94 (2004)

Author Index

Printed in the United States
By Bookmasters